AF581704

Recent Advances in Post-Lithium Ion Batteries

Recent Advances in Post-Lithium Ion Batteries

Editor

Eliana Quartarone

MDPI • Basel • Beijing • Wuhan • Barcelona • Belgrade • Manchester • Tokyo • Cluj • Tianjin

Editor
Eliana Quartarone
University of Pavia
Italy

Editorial Office
MDPI
St. Alban-Anlage 66
4052 Basel, Switzerland

This is a reprint of articles from the Special Issue published online in the open access journal *Batteries* (ISSN 2313-0105) (available at: https://www.mdpi.com/journal/batteries/special_issues/Recent_Advances_Post-Lithium_Ion_Batteries).

For citation purposes, cite each article independently as indicated on the article page online and as indicated below:

LastName, A.A.; LastName, B.B.; LastName, C.C. Article Title. *Journal Name* **Year**, *Volume Number*, Page Range.

ISBN 978-3-0365-1244-0 (Hbk)
ISBN 978-3-0365-1245-7 (PDF)

Contents

About the Editor

Eliana Quartarone is an associate professor of electrochemistry and physical chemistry at the Department of Chemistry of the University of Pavia. Her research field is focused on the development of advanced materials (synthesis and characterization) for energetics, electrochemistry, and, specifically, Li batteries, PEMFC, and energy harvesting.

Preface to "Recent Advances in Post-Lithium Ion Batteries"

Lithium batteries are efficient storage systems for portable electronic devices, electrical power grids, and electrified transportation due to their high-energy density and low maintenance requirements. After their launch into the market in 1990s, they immediately became the dominant technology for portable systems. The development of lithium ion batteries (LiBs) for electric drive vehicles has been, in contrast, rather incremental. There are several critical issues, such as an energy density, system safety, cost, and environmental impact of the battery production processes, that remain challenges, especially in the automotive field. All of these concerns bring into question the suitability of the LiB to satisfy the ever-growing energy requirements of the long-term future. In order to strengthen the LiB's competitiveness and affordability in vehicle technology, the necessity of game-changer batteries is urgent. Recently, a novel approach going beyond Li-ion batteries has become rapidly established. Several new chemistries and strategies have been proposed, leading to promising performances in terms of energy density, long-life storage capability, safety, and sustainability. However, several challenges, such as a thorough understanding of mechanisms, cell design, long-term durability, and safety issues, have not yet been fully addressed. This book includes six high-quality papers reporting some recent developments and emerging trends in the field of "post-lithium" batteries. The first article deals with the use of EIS in the investigation of oxygen evolution reaction (OER) and oxygen reduction reaction (ORR) kinetics in Li-O2 batteries using palladium-based catalysts. Two different families of Li-ion-conducting glass-forming polymer electrolytes are also described as novel solid electrolytes for all-solid-state Li-ion batteries, discussing the relationship between transport properties and polymer dynamics. Another paper reports on the influence of heating temperature and time on the interfacial resistance between a ceramic electrolyte and the Li metal electrode in order to evaluate the stability of lithium electrodeposition/dissolution. The fifth article describes the synthesis and characterization of Na0.44MnO2 nanometric slabs, showing improved capacities at higher rates due to a lower degree of anisotropy. Finally, microscopic and macroscopic modelling methods are discussed in the case of metal–air batteries, with a focus on continuum modelling derived from non-equilibrium thermodynamics. This book will be beneficial for researchers and engineers involved in the field of material science and batteries.

Eliana Quartarone
Editor

Article

Mechanism of Ionic Impedance Growth for Palladium-Containing CNT Electrodes in Lithium-Oxygen Battery Electrodes and Its Contribution to Battery Failure

Neha Chawla [1], Amir Chamaani [2], Meer Safa [1], Marcus Herndon [1] and Bilal El-Zahab [1,*]

1 Department of Mechanical and Materials Engineering, Florida International University, Miami, FL 33174, USA; nchaw002@fiu.edu (N.C.); msafa@fiu.edu (M.S.); mhern544@fiu.edu (M.H.)

2 Department of Mechanical and Industrial Engineering, The University of Illinois at Chicago, Chicago, IL 60607, USA; acham027@fiu.edu

* Correspondence: belzahab@fiu.edu; Tel.: +1-305-348-3558

Received: 30 November 2018; Accepted: 22 January 2019; Published: 23 January 2019

Abstract: The electrochemical oxygen evolution reaction (OER) and oxygen reduction reaction (ORR) and on CNT (carbon nanotube) cathode with a palladium catalyst, palladium-coated CNT (PC-CNT), and palladium-filled CNT (PF-CNT) are assessed in an ether-based electrolyte solution in order to fabricate a lithium-oxygen battery with high specific energy. The electrochemical properties of the CNT cathodes were studied using electrochemical impedance spectroscopy (EIS). Palladium-filled cathodes displayed better performance as compared to the palladium-coated ones due to the shielding of the catalysts. The mechanism of the improvement was associated to the reduction of the rate of resistances growth in the batteries, especially the ionic resistances in the electrolyte and electrodes. The scanning electron microscopy (SEM) and spectroscopy were used to analyze the products of the reaction that were adsorbed on the electrode surface of the battery, which was fabricated using palladium-coated and palladium-filled CNTs as cathodes and an ether-based electrolyte.

Keywords: EIS; Fourier-Transform Infrared Spectroscopy; cycling; catalyst; carbon nanotubes; $Li\text{-}O_2$ battery

---

1. Introduction

Rechargeable batteries convert chemical energy to electrical energy by the reversible electrochemical reactions of reduction and oxidation. Various types of secondary battery systems that have been studied, developed, and used in the past, include nickel-metal hydride, lead-acid, lithium-ion batteries, and redox-flow batteries [1–4]. Owning to their high specific energy, lithium-ion batteries have found applications in various fields. Due to the growing demand for advanced energy storage solutions for the smart grids, automotive industries, and other consumer applications, the research for ultra-high theoretical specific energy has led to studies beyond lithium-ion batteries, where metal-air batteries have been at the forefront of this research [5–7]. Metal-air batteries are unique because the cathode material, i.e., oxygen gas, can be obtained from the atmosphere, instead of storing it in the battery. Lithium–air batteries have a high theoretical specific energy density of 3500 $Wh.kg^{-1}$ (considering the cathode), which is many folds higher than current lithium-ion batteries [8,9].

Abraham et al. first reported on a non-aqueous lithium-oxygen battery in 1996, which consisted of a lithium anode, a polymer electrolyte, and a porous carbon-air cathode [9]. Since then, several studies have been conducted to study the rechargeability of $Li\text{-}O_2$ batteries. During the discharging of a $Li\text{-}O_2$ battery, Li^+ migrates through the electrolyte to the cathode, where the incoming electrons from the external circuit combine with oxygen from the atmosphere to form peroxide ions, O_2^{2-}, which, in turn, reacts with the Li^+ in an oxygen reduction reaction (ORR) to form the discharge product, Li_2O_2. During charging, a reverse reaction, as well as an oxygen evolution reaction (OER) occurs, regenerating the Li^+ and O_2. The reaction is represented as follows:

$$2Li^+ + i\,O_2 + 2e^- \leftrightarrow Li_2O_2.$$

Many carbonaceous materials have been used as cathodes for Li-O_2 batteries [10]. Some of the cathode materials used have been carbon nanoparticles [11,12], carbon nanofibers [13,14], carbon nanotubes [10,12,13,15–17], graphene platelets [18,19], and other forms of carbons [20,21]. Carbon nanotubes (CNTs) are known to have high electrical conductivity, a high specific area, more numbers of accessible active sites for reactions, and better chemical stability. All these properties make them a suitable choice for use as Li-O_2 battery cathodes [10,17,22–24].

Several metal and metal-oxide catalysts have been used in and as cathodes of Li-O_2 batteries in order to improve the catalytic properties of the OER and ORR [25–29]. Some of the elements used have been platinum [30], palladium [31], ruthenium [26,27,32–35], gold [30,36], and various metal oxides [37–39]. The presence of a catalyst destabilizes the oxidizing species, which results in decreased charging overpotential in Li-O_2 batteries [40–42]. It also improves the cyclability [24–27,40], and thus the overall battery performance. Gittleson et al. [43,44] have shown that palladium and platinum catalysts promote Li_2O_2 oxidation at lower potentials. However, they also reported the occurrence of electrolyte decomposition that resulted in the formation of lithium carbonates, which deactivated the catalysts. The porous carbon cathode was passivated by a protective Al_2O_3 layer, applied using atomic layer deposition by Khalil et al. [21] prior to coating Pd catalysts. Their approach led to the formation of a nanocrystalline Li_2O_2 during discharge that helped improve the electronic transport properties, thus reducing the charge over-potential to about 0.2 V [21]. According to Li et al., Ru nanocrystals, when used as catalysts, decomposed Li_2CO_3 at 3.5 V, thus reducing the accumulation of Li_2CO_3 in in the cathode during cycling.

In order to overcome the agglomeration of catalysts and deactivation of Li-O_2 batteries, dispersing catalysts were dispersed in polymeric membranes over the oxygen electrode, and an atomic layer deposition was used to chemically bind the catalysts to the surface of carbon nanotubes, where the CNTs were then filled with catalysts [10,35,45–48]. In our previous work, we filled the CNT annulus with palladium nanocatalysts and reported increased stability of the electrolyte during the charge and discharge process [10]. In addition, the full discharge of the battery to 2 V resulted in a six-fold increase in the first discharge capacity compared to the pristine CNT cathodes, and over 35% for their PC-cathodes.

In this paper, we investigate and report the electrochemical impedance performance of PC-and PF-CNT cathodes during charge/discharge cycling to unveil the failure mechanism of the batteries. We also investigate the chemical nature and morphology of deposits on the electrodes using FTIR (Fourier Transform Infrared Spectroscopy) and scanning electron microscopy (SEM) to corroborate the results from the electrochemical impedance spectroscopy (EIS) study.

2. Experimental

Materials: Bis (trifluoromethane) sulfonamide with a purity of >99.8% (LiTFSI), tetraethylene glycol dimethyl ether of >99.00% purity (TEGDME), N-Methyl pyrrolidine of >97% purity, and multi-walled carbon nanotubes (MWCNTs) with a diameter of 5–20 nm, length of 5 μm, and purity >96% of carbon basis were purchased from Sigma Aldrich. Palladium (II) chloride was purchased from ACROS organics. PVDF (Polyvinylidene fluoride) was procured from Alfa Aesar. Fuel Cell Earth supplied the carbon cloth gas diffusion layer (CCGDL) with a thickness of ~300 μm. Lithium chips with a purity >99.90% and a Celgard polypropylene separator with a thickness of ~25 μm were bought from MTI Corp. Scheme 1 depicts the assembly of the batteries.

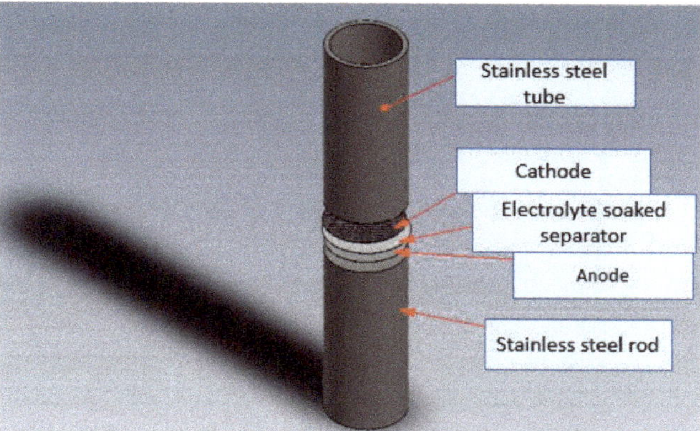

Scheme 1. Schematic depiction of the assembled Li-O_2 batteries.

Preparation of the Electrode: We have followed the procedure for PF-CNTs preparation as described in our previous paper [10,49]. Briefly, nitric acid treatment was performed on the MWCNTs to decap the ends. 100 mg of decapped multi-walled carbon nanotubes (MWCNTs) was blended with 1 mM aqueous solution of $PdCl_2$ until a slurry was formed. A similar procedure was followed on the untreated capped MWCNTs to obtain PC-CNTs. Both slurries were preserved at 25 °C for 10 h and then calcinated in air at 350 °C for ~2 h. Calcination was followed by hydrogenation that resulted in a yield of ~5 wt.% palladium nanoparticles. Preparation of cathode was as mentioned in the preceding paper [10]. Slurry of PVDF/PF and PC-CNT (10/90 wt.% in N-Methyl-2-pyrrolidone or NMP) was used to coat the cathodes on carbon cloth with a 0.5″ diameter. The cathodes were dried for 12 hours at 120 °C and then stocked in an argon-filled glovebox. The loading of MWCNT was 0.39 ± 0.01 mg.cm^{-2}. All the capacities reported in this manuscript are per total mass of active cathode including the catalyst and CNTs.

Electrolyte Preparation and Battery Test Assembly: 1 M LiTFSI in TEGDME solution was used to prepare the electrolyte. Modified Swagelok-type cells were used for our battery assemblies. It consisted of a stainless-steel tube on the cathode side and a stainless-steel rod on the anode side. Celgard 2400 separator which was soaked with the electrolyte was arranged on the lithium metal disc anode. Then, the MWCNT-CCGDL was placed on it, followed by a stainless-steel mesh current collector. The batteries were then placed inside an Ar-filled glove box (<1 ppm O_2 and <0.1 ppm H_2O) for resting overnight before the electrochemical tests.

Electrochemical Characterizations: Before testing the batteries, they were rested under oxygen for 5 h. Galvanostatic discharge/charge tests were performed on the Gamry 600^{TM} Reference electrochemical workstation. Charge/discharge tests were performed within 2.0–4.5 V voltage range at a 250 mA·g^{-1} current density. The electrochemical characteristics of the cell were investigated by Electrochemical Impedance Spectroscopy (EIS) performed using a Gamry 600^{TM} electrochemical workstation. Impedance data was collected in the frequency range of 10^6–0.01 Hz. The environment was temperature-controlled at 25 °C for all charge/discharge and electrochemical tests. After cycling, the oxygen cathodes were recovered, rinsed with dimethyl carbonate (DMC), and dried under vacuum inside the Ar-filled glove box.

Infrared Characterizations: The cathodes were investigated by FTIR (Fourier transform infrared) spectroscopy using JASCO FT-IR 4100 (company, city, country) for product characterization in the 500–4000 cm^{-1}, mid-IR frequency region. A resolution of 4 cm^{-1} was used for each spectrum. For pretreatment for FTIR, the samples were rinsed three times with DMC (dimethyl carbonate) to remove the electrolyte salt from the cathode surface, and dried in vacuum to volatize the DMC.

3. Results and Discussion

Li-O_2 batteries with cathodes comprising PC and PF-CNTs showed full discharge capacities of ~8200 mAh.g^{-1} and ~11,200 mAh.g^{-1}, respectively [10]. The batteries were discharged to 2 V under oxygen. Galvanostatic cycling was employed to characterize the electrochemical properties on the cathodes during the OER and ORR processes. Cycle performance of the Li-O_2 batteries with PC-and PF-CNTs as cathodes were investigated at fixed capacities of 250 mAh.g^{-1}, 500 mAh.g^{-1}, and 1000 mAh.g^{-1} at a 250 mA.g^{-1} current density between 2.0–4.5 V (Figure 1). At 250 mAh.g^{-1}, 41 and 81 cycles were obtained using PC-and PF-CNTs, respectively. At 500 mAh.g^{-1}, 44 and 58 cycles were obtained using PC-and PF-CNTs, respectively. 16 and 25 cycles were obtained at 1000 mAh.g^{-1}. Figure 2 shows the terminal discharge voltages of the batteries.

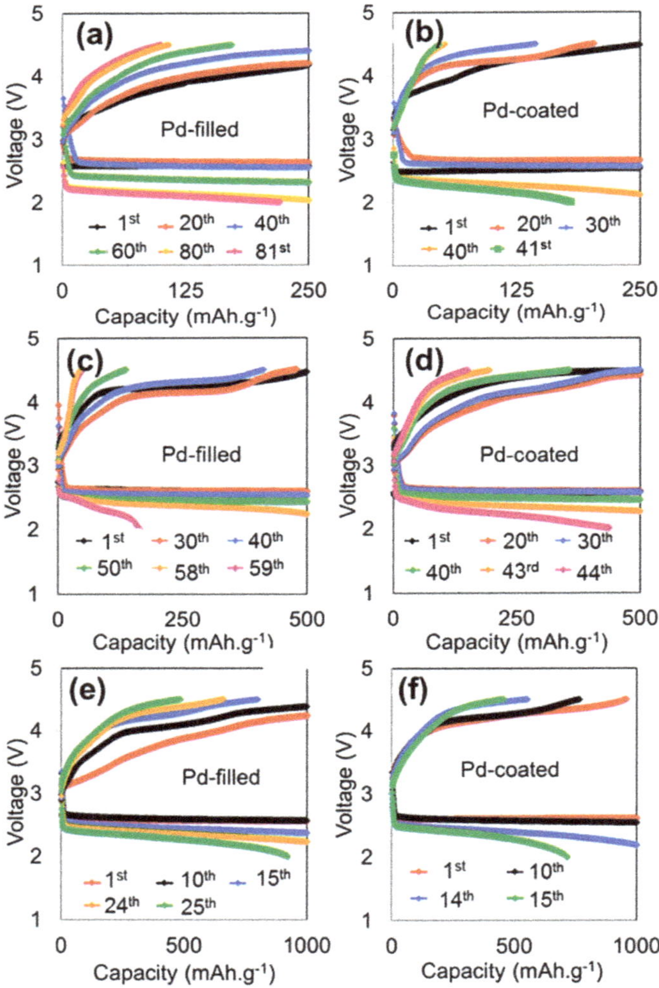

Figure 1. Voltage profile of lithium-oxygen (Li−O_2) batteries with (**a**,**c**,**e**) palladium-filled and (**b**,**d**,**f**) palladium-coated carbon nanotubes (CNTs) at fixed capacities of 250 mAh.g^{-1}, 500 mAh.g^{-1}, and 1000 mAh.g^{-1}, respectively, at a 250 mA.g^{-1} current density between 2.0 to 4.5 V.

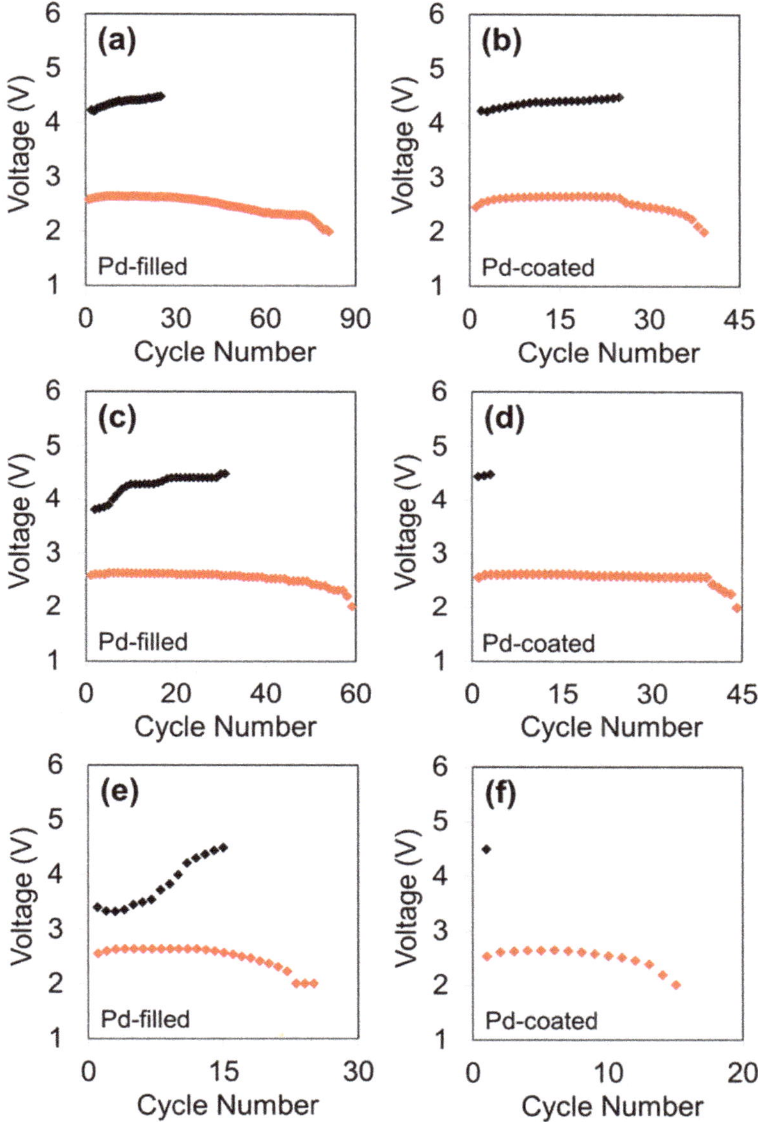

Figure 2. End charge/discharge voltages of (**a**,**c**,**e**) palladium-filled and (**b**,**d**,**f**) palladium-coated CNTs after full cycles to 250 mAh.g^{-1}, 500 mAh.g^{-1}, and 1000 mAh.g^{-1}, respectively.

EIS measurements were performed after discharge/charge cycles to investigate the cathode/electrolyte behavior during discharging and charging processes. The Nyquist plot after discharge shows a high-frequency intercept, R_b, which correlates to the resistance of cell, including contributions from the electrolyte, electrodes, and contact resistances [16,50–56], as shown in Figure 3g. R_b is the high-frequency intercept and is obtained from the real EIS spectra. The EIS spectrum is followed by the semicircle that is associated with charge-transfer resistance at the cathode/O_2/electrolyte interface [50–52], represented by R_{int} (interfacial resistance). R_{int} is obtained from the diameter of the semicircle of EIS. This is generally attributed to the Li_2O_2 which forms on the surface of the cathode after discharge, which hinders charge transfer and O_2 diffusion through the pores of the cathode [51]. Following R_{int} there is a linear rise which is known as the Warburg-like linear region, which has also been reported in other metal-O_2 battery EIS spectra at non-Faradaic conditions. In order to determine the impedance behavior of the porous electrodes in lithium-ion and metal−O_2 batteries, the TLM or transmission line model was used. It is assumed that the Warburg-like linear region corresponds to the resistance of lithium-ion migration at the cathode, and is represented as R_{ion}. We used a TLM model and estimated the R_{ion} by the projection of the Warburg-like line on the impedance abscissa (R_{ion}/3), as shown in Figure 3b [16,54].

In Figure 3, the Nyquist plot exhibits a semicircle with a smaller diameter for PF-CNT, and a semicircle with a bigger diameter for the PC-CNTs after the first discharge to 250 mAh.g^{-1}. The semicircle diameter corresponds to the charge-transfer resistance [51]. Higher charge-transfer resistance leads to slower charge-transfer kinetics [51]. From Figure 4, it is evident that the PC-cathode has slower charge-transfer kinetics as compared to PF-CNTs, due to higher resistance. As the number of cycles increase, we can see a shift of the high-frequency intercept, R_b, toward the right. From Figure 3a,b and Figure 4a,b, the shift is clearly visible for both cathodes. Figure 4 shows the change in the resistances after discharging the batteries to 250 mAh.g^{-1}(a,b), 500 mAh.g^{-1}(c,d), and 1000 mAh.g^{-1} (e,f). After the last discharging cycle up to 250 mAh.g^{-1}, which is 41 cycles for PC-CNT and 81 cycles of PF-CNT, respectively, Rb has shifted from 80 Ω to 2910 Ω for PC-CNT, whereas the shift of R_b for PF-CNT is from 80 Ω to 300 Ω (Figure 4a,b). This shift in R_b is attributed to electrolyte decomposition and formation of lithium carbonates over time. In Figure 4c, for PF-CNT discharged up to 500 mAh.g^{-1}, the R_b increases from 70 Ω for the first cycle, to 300 Ω for the last cycle. For PC-CNT (Figure 4d), the R_b increases from 90 Ω to 1700 Ω. In Figure 4e, for PF-CNT discharged up to 1000 mAh.g^{-1}, the R_b increases from 70 Ω for the first cycle to 250 Ω for the 25th cycle. However, in the case of PC-CNT (Figure 4f), the R_b increases from 80 Ω to 1200 Ω. The gradual increase in the value of R_b is credited to the slow accumulation of the reaction products, Li_2O_2 and Li_2O, within the pores of the carbon electrode, whereas the steep increase is an indication of a complete blockage of the carbon electrode that prevents further reduction of O_2 [56]. This shift in R_b is attributed to electrolyte decomposition and formation of lithium carbonate (Li_2CO_3) over time [51]. As seen in Figure 4, R_{int} decreases minimally for the first 30 cycles of PF-CNTs and then increases. For PC-CNTs, R_{int} decreases for the first 10 cycles after the first discharge and then keeps increasing. R_{ion} was the major resistance in both cathodes, and it keeps increasing with the number of cycles, as seen in the figure. Anode/electrolyte interface initially governs the interfacial resistance of Li-O_2 batteries [24,57]. The initial decrease in the value of R_{int} is mainly attributed to the dissolution of the passivation film on the anode/electrolyte film's interface [16]. Accumulation of irreversible discharge/charge by-products on the anode/electrolyte [16] and cathode/electrolyte interfaces [16,57] results in an increased R_{int} value during later cycles. Yi et al. [58,59] showed that an increased interfacial resistance on the cathode/electrolyte interface in Li-O_2 batteries could be the result of lithium carbonate by-product formation during cycling. According to Shui et al. [60], deactivation of the cathode leads to Li-O_2 battery death. Increase in R_{ion} during cycling is an indication that the cathode pores have been clogged, meaning that Li^+ ion transport is hindered [16].

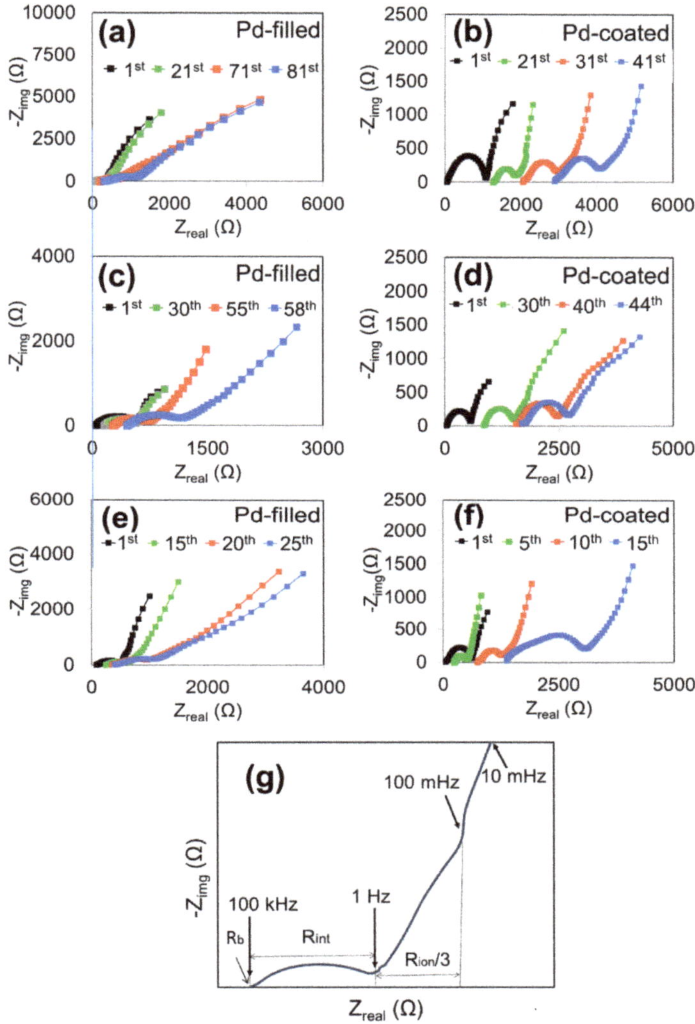

Figure 3. Nyquist plots of (**a**) PF- and (**b**) PC-CNTs after different discharging cycles of 250 mAh.g^{-1} capacity; Nyquist plots of (**c**) PF- and (**d**) PC-CNTs after different discharging cycles of 500 mAh.g^{-1} capacity; and Nyquist plots of (**e**) PF- and (**f**) PC-CNTs after different discharging cycles of 1000 mAh.g^{-1} capacity. (**g**) Typical Nyquist plot. The batteries were cycled at 250 mA.g^{-1} with a current density between 2 to 4.5 V.

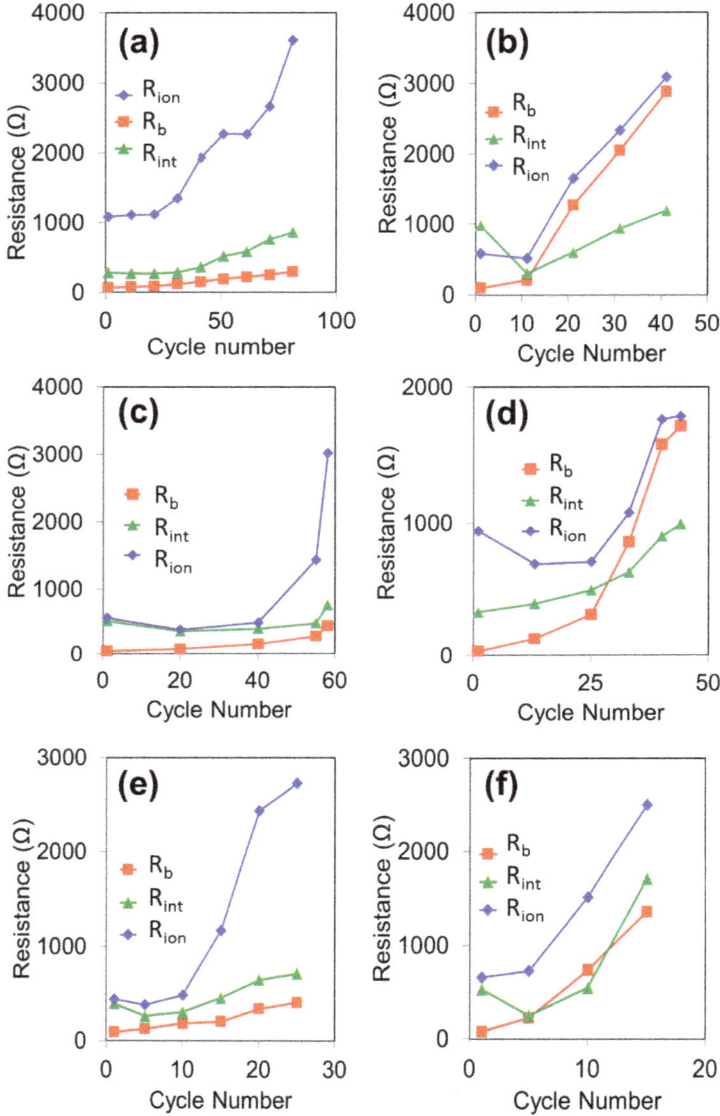

Figure 4. Change of resistances after the discharging (**a**,**c**,**e**) Palladium-filled CNT, and (**b**,**d**,**f**) palladium-coated CNT cycles of 250 mAh.g^{-1}, 500 mAh.g^{-1}, and 1000 mAh.g^{-1} capacities, respectively.

For the cycles of 1000 mAh.g^{-1} as the fixed capacity, similar Nyquist plots were observed (Figure 4c,d). In this case too, the PF-CNT was cycled for 25 cycles, as compared to 16 cycles with PC-CNTs. After the last discharging cycle of up to 1000 mAh.g^{-1}, R_b shifted from 80 to 1350 for PC-CNT, whereas the shift for PF-CNT was from 80 to 400. This indicates more electrolyte decomposition in the case of PC-CNTs. R_{ion} had the highest resistance for both cases. Hence, in this case too, the battery failure was primarily due to pore-clogging.

As shown by the EIS, the formation of Li_2O_2 hinders the charge transfer between the interfaces and the decomposition of electrolyte [16]. Both reasons lead to the failure of Li-O_2 batteries. To investigate the composition of the deposits that clog the pores of the cathodes, we performed FTIR on PC and PF-CNT cathodes after cycling for 15 and 45 cycles at fixed capacities of 1000 mAh.g^{-1} and 250 mAh.g^{-1}. The peak at around 590 cm^{-1} is attributed to Li_2O_2, and Li_2CO_3 produces a peak at 862 cm^{-1} [61]. From Figure 5, it can be observed that the peaks of Li_2CO_3 are higher in the case of PC, rather than PF-CNTs. Better cycling was observed in PF-CNTs due to less undesirable products. The improvement in the cycling stability of PF-CNTs over the PC-CNTs was a result of the decrease in undesirable charge/discharge product formation, e.g., Li_2CO_3, caused by the encapsulation approach [10].

Figure 6 shows a comparison between PC-and PF-CNTs at capacities of 500, 1000, 1500, 5000, and 8000 mAh.g^{-1}. As observed from the figure, at 500 mAh.g^{-1}, a Li_2O_2 peak is present in both cathodes. A distinct Li_2CO_3 peak is observed for PC-CNT but is negligible in the case of PF-CNT. At all the capacities, Li_2O_2 peaks are present for both PC-and PF-CNT cathodes. However, the Li_2CO_3 peak for PF-CNT cathodes gradually increases as the capacity increases. Low Li_2CO_3 formation in the PF-CNT, as compared to PC-CNT results in increased battery performance [61].

SEM proved the formation of Li_2CO_3 dendrites on the cathode surface after cycling, as shown in Figure 7. When the batteries were cycled between 2.0 and 4.0 V, there is a constant accumulation of lithium carbonate (Li_2CO_3) arising from the reactions involving the electrode and electrolyte [62], which results in electrode passivation and capacity fading in cells with carbon electrodes. Lithium carbonates are also produced as a result of the side-reaction of lithium peroxide with moisture and carbon dioxide from the atmosphere. It is difficult to be decomposed, and considerably deteriorated the cycling life [20,63]. The large over-charge potential can induce the side-reaction of oxidation decomposition of the electrode and electrolyte during Li_2O_2 oxidation, which results in the formation of Li_2CO_3 [62]. The accumulation of Li_2CO_3 in the electrode led to electrode passivation and capacity-fading on cycling. The PC-CNT cathode shown in Figure 7b presents platelet-shaped lithium peroxide (Li_2O_2) covered by a thick conformal layer of lithium carbonates (Li_2CO_3), as shown in the FTIR measurements whereas the PF-CNTs showed nano-thin platelets of Li_2O_2 canvasing the cathode (Figure 7a). Since the Li_2O_2 platelets yield high surface porosity, they do not block access to the CNTs which in turn enhances battery performance [10].

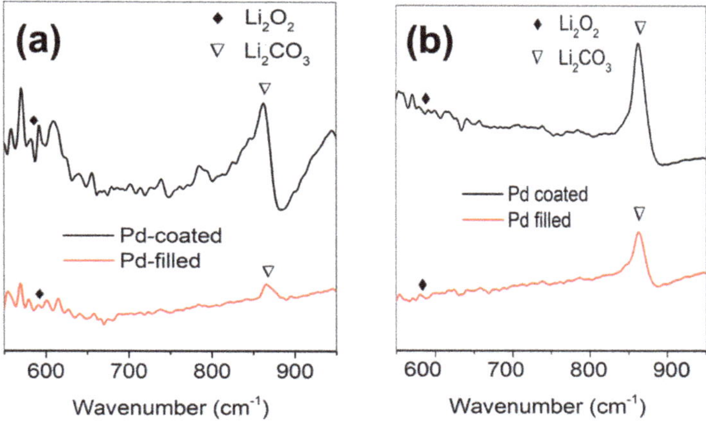

Figure 5. Fourier transform infrared (FTIR) spectra of PF- and PC-CNTs after cycling for (**a**) 45 cycles to 250 mAh.g^{-1} and (**b**) 15 cycles to 1000 mAh.g^{-1}.

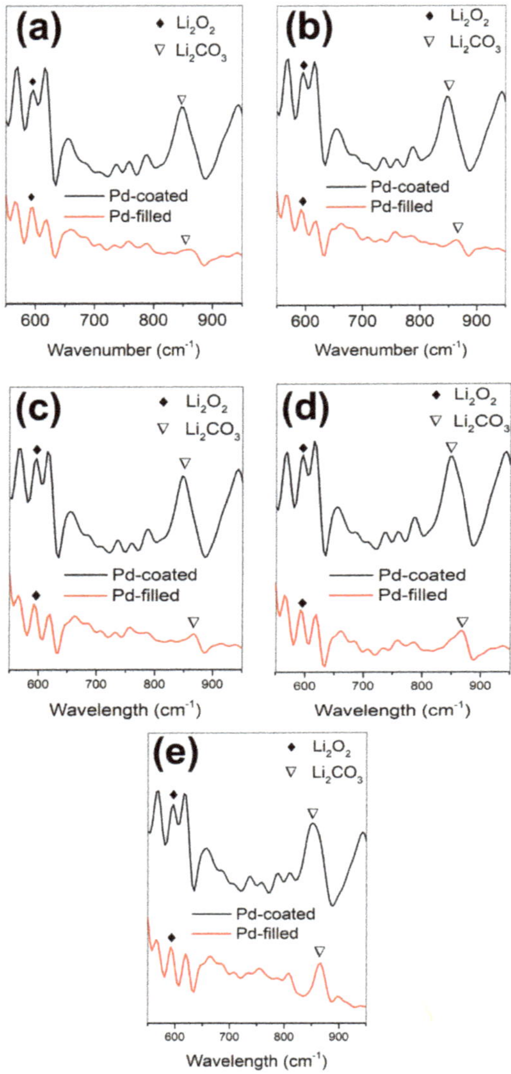

Figure 6. FTIR spectra after discharging to (**a**) 500 $mAh.g^{-1}$, (**b**) 1000 $mAh.g^{-1}$, (**c**) 1500 $mAh.g^{-1}$, (**d**) 5000 $mAh.g^{-1}$, and (**e**) 8000 $mAh.g^{-1}$. All batteries were discharged at 250 $mA.g^{-1}$ between 2 to 4.5 V.

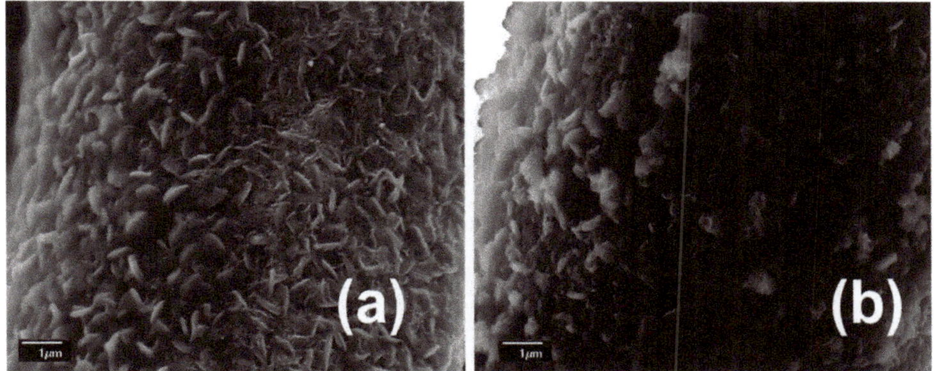

Figure 7. SEM images after cycling of (**a**) PF- and (**b**) PC-CNT cathodes.

As we demonstrated in our previous paper [10], the improvement in cycling stability of PF- CNTs over the PC-CNTs was a result of the decrease in undesirable charge/discharge product formation, e.g., Li_2CO_3, afforded by the encapsulation approach.

4. Conclusions

An EIS study was conducted in this paper on a Li-O_2 battery utilizing PC-and PF- CNTs, and we provided insight into kinetics operating the battery operation. From EIS results, it was shown that R_{ion} had the highest resistance, which indicates the failure of the batteries due to pore-clogging of the cathodes. Increasing R_b resistamce indicates more electrolyte decomposition in the case of PC, rather than PF-, CNT cathodes. FTIR and SEM further confirm that PF-CNTs have better capacity-retention due to less Li_2CO_3 formation on the surface of the cathode, as compared to PC-CNTs.

Author Contributions: Investigation, N.C., A.C., M.S. and M.H.; Supervision, B.E.-Z.

Funding: This research received no external funding.

Acknowledgments: The authors acknowledge Peggy Wang's group for FTIR instrument use. N.C. and M.S. acknowledge Florida International University Dissertation Year Fellowship for the financial support. M.H. acknowledges the financial support of the McKnight Doctoral Fellowship.

Conflicts of Interest: The authors declare no conflict of interest.

References

1. Feng, N.; He, P.; Zhou, H. Critical Challenges in Rechargeable Aprotic Li-O_2 Batteries. *Adv. Energy Mater.* **2016**, *6*, 1–24. [CrossRef]
2. Panchal, S.; Mathewson, S.; Fraser, R.; Culham, R.; Fowler, M. Thermal Management of Lithium-Ion Pouch Cell with Indirect Liquid Cooling using Dual Cold Plates Approach. *SAE Int. J. Altern. Powertrains* **2015**, *4*, 293–307. [CrossRef]
3. Safa, M.; Chamaani, A.; Chawla, N.; El-Zahab, B. Polymeric Ionic Liquid Gel Electrolyte for Room Temperature Lithium Battery Applications. *Electrochim. Acta* **2016**, *213*, 587–593. [CrossRef]
4. Panchal, S.; Mathewson, S.; Fraser, R. *Experimental Measurements of Thermal Characteristics of LiFePO4 Battery*; SAE Tech. Paper; SAE: Warrendale, PA, USA, 2015.
5. Geng, D.; Ding, N.; Hor, T.S.A.; Chien, S.W.; Liu, Z.; Wuu, D.; Sun, X.; Zong, Y. From Lithium-Oxygen to Lithium–air Batteries: Challenges and Opportunities. *Adv. Energy Mater.* **2016**, 1–14. [CrossRef]
6. Landa-Medrano, I.; Li, C.; Ortiz-Vitoriano, N.; de Larramendi, I.R.; Carrasco, J.; Rojo, T. Sodium-Oxygen Battery: Steps Toward Reality. *J. Phys. Chem. Lett.* **2016**, *7*, 1161–1166. [CrossRef] [PubMed]

7. Safa, M.; Hao, Y.; Chamaani, A.; Adelowo, E.; Chawla, N.; Wang, C.; El-Zahab, B. Capacity Fading Mechanism in Lithium-Sulfur Battery using Poly(ionic liquid) Gel Electrolyte. *Electrochim. Acta* **2017**, *258*, 1284–1292. [CrossRef]
8. Girishkumar, G.; McCloskey, B.; Luntz, A.C.; Swanson, S.; Wilcke, W. Lithium–air battery: Promise and challenges. *J. Phys. Chem. Lett.* **2010**, *1*, 2193–2203. [CrossRef]
9. Abraham, K.M.; Jiang, Z. A Polymer Electrolyte—Based Rechargeable Lithium/Oxygen Battery technical papers electrochemical science and technology A Polymer Electrolyte-Based Rechargeable lithium/Oxygen Battery. *J. Electrochem. Soc.* **1996**, *143*, 1–5. [CrossRef]
10. Chawla, N.; Chamaani, A.; Safa, M.; El-Zahab, B. PF- Carbon Nanotubes Cathode for Improved Electrolyte Stability and Cyclability Performance of Li-O_2 Batteries. *J. Electrochem. Soc.* **2017**, *164*, A6303–A6307. [CrossRef]
11. Zhang, Y.; Zhang, H.; Li, J.; Wang, M.; Nie, H.; Zhang, F. The use of mixed carbon materials with improved oxygen transport in a Lithium–air battery. *J. Power Sources* **2013**, *240*, 390–396. [CrossRef]
12. Fuentes, R.E.; Colón-Mercado, H.R.; Fox, E.B. Electrochemical evaluation of carbon nanotubes and carbon black for the cathode of Li-air batteries. *J. Power Sources* **2014**, *255*, 219–222. [CrossRef]
13. Zhang, G.Q.; Zheng, J.P.; Liang, R.; Zhang, C.; Wang, B.; Hendrickson, M.; Plichta, E.J. Lithium–Air Batteries Using SWNT/CNF Buckypapers as Air Electrodes. *J. Electrochem. Soc.* **2010**, *157*, A953. [CrossRef]
14. Mitchell, R.R.; Gallant, B.M.; Thompson, C.V.; Shao-Horn, Y. All-carbon-nanofiber electrodes for high-energy rechargeable Li-O_2 batteries. *Energy Environ. Sci.* **2011**, *4*, 2952. [CrossRef]
15. Li, Y.; Wang, J.; Li, X.; Liu, J.; Geng, D.; Yang, J.; Li, R.; Sun, X. Nitrogen-doped carbon nanotubes as cathode for lithium–air batteries. *Electrochem. Commun.* **2011**, *13*, 668–672. [CrossRef]
16. Chamaani, A.; Safa, M.; Chawla, N.; El-Zahab, B. Composite Gel polymer electrolyte for improved cyclability in lithium-oxygen batteries. *ACS Appl. Mater. Interfaces* **2017**, *9*, 33819–33826. [CrossRef]
17. Chamaani, A.; Chawla, N.; Safa, M.; El-Zahab, B. One-Dimensional Glass Micro-Fillers in Gel Polymer Electrolytes for Li-O_2 Battery Applications. *Electrochim. Acta* **2017**, *235*, 56–63. [CrossRef]
18. Li, Y.; Wang, J.; Li, X.; Geng, D.; Banis, M.N.; Li, R.; Sun, X. Nitrogen-doped graphene nanosheets as cathode materials with excellent electrocatalytic activity for high capacity lithium-oxygen batteries. *Electrochem. Commun.* **2012**, *18*, 12–15. [CrossRef]
19. Zu, C.; Li, L.; Qie, L.; Manthiram, A. Expandable-graphite-derived graphene for next-generation battery chemistries. *J. Power Sources* **2015**, *284*, 60–67. [CrossRef]
20. Zhang, T.; Zhou, H. A reversible long-life Lithium–air battery in ambient air. *Nat. Commun.* **2013**, *4*, 1817. [CrossRef]
21. Lu, J.; Lei, Y.; Lau, K.C.; Luo, X.; Du, P.; Wen, J.; Assary, R.S.; Das, U.; Miller, D.J.; Elam, J.W.; et al. A nanostructured cathode architecture for low charge overpotential in lithium-oxygen batteries. *Nat. Commun.* **2013**, *4*, 2383. [CrossRef]
22. Astinchap, B.; Moradian, R.; Ardu, A.; Cannas, C.; Varvaro, G.; Capobianchi, A. Bifunctional FePt @ MWCNTs/Ru Nanoarchitectures: Synthesis and Characterization. *Chem. Mater.* **2012**, *24*, 3393–3400. [CrossRef]
23. Itkis, D.M.; Semenenko, D.A.; Kataev, E.Y.; Belova, A.I.; Neudachina, V.S.; Sirotina, A.P.; Hävecker, M.; Teschner, D.; Knop-Gericke, A.; Dudin, P.; et al. Reactivity of carbon in lithium-oxygen battery positive electrodes. *Nano Lett.* **2013**, *13*, 4697–4701. [CrossRef] [PubMed]
24. Chamaani, A.; Safa, M.; Chawla, N.; Herndon, M.; El-zahab, B. Stabilizing e ff ect of ion complex formation in lithium—Oxygen battery electrolytes. *J. Electroanal. Chem.* **2018**, *815*, 143–150. [CrossRef]
25. Rahman, M.A.; Wang, X.; Wen, C. A review of high energy density lithium–air battery technology. *J. Appl. Electrochem.* **2013**, *44*, 5–22. [CrossRef]
26. Yilmaz, E.; Yogi, C.; Yamanaka, K.; Ohta, T.; Byon, H.R. Promoting Formation of Noncrystalline Li_2O_2 in the Li-O_2 Battery with RuO_2 Nanoparticles. *Nano Lett.* **2013**, *13*, 4679–4684. [CrossRef] [PubMed]
27. Sun, B.; Munroe, P.; Wang, G. Ruthenium nanocrystals as cathode catalysts for lithium-oxygen batteries with a superior performance. *Sci. Rep.* **2013**, *3*, 1–7. [CrossRef] [PubMed]
28. Débart, A.; Paterson, A.J.; Bao, J.; Bruce, P.G. α-MnO_2 Nanowires: A Catalyst for the O_2 Electrode in Rechargeable Lithium Batteries. *Angew. Chem. Int. Ed.* **2008**, *47*, 4521–4524. [CrossRef]
29. Chawla, N. Nanocatalysts in lithium oxygen batteries. *Nano Sci. Nano Technol. Indian J.* **2018**, *12*, 128.

30. Lu, Y.C.; Gasteiger, H.a.; Shao-Horn, Y. Catalytic activity trends of oxygen reduction reaction for nonaqueous Li-air batteries. *J. Am. Chem. Soc.* **2011**, *133*, 19048–19051. [CrossRef]
31. Xu, J.-J.; Wang, Z.-L.; Xu, D.; Zhang, L.-L.; Zhang, X.-B. Tailoring deposition and morphology of discharge products towards high-rate and long-life lithium-oxygen batteries. *Nat. Commun.* **2013**, *4*, 2438. [CrossRef]
32. Zhao, G.; Niu, Y.; Zhang, L.; Sun, K. Ruthenium oxide modified titanium dioxide nanotube arrays as carbon and binder free lithium–air battery cathode catalyst. *J. Power Sources* **2014**, *270*, 386–390. [CrossRef]
33. Wang, Y.; Rong, Z.; Wang, Y.; Qu, J. Ruthenium nanoparticles loaded on functionalized graphene for liquid-phase hydrogenation of fine chemicals: Comparison with carbon nanotube. *J. Catal.* **2016**, *333*, 8–16. [CrossRef]
34. Guo, Z.; Zhou, D.; Liu, H.; Dong, X.; Yuan, S.; Yu, A.; Wang, Y.; Xia, Y. Synthesis of ruthenium oxide coated ordered mesoporous carbon nanofiber arrays as a catalyst for lithium oxygen battery. *J. Power Sources* **2015**, *276*, 181–188. [CrossRef]
35. Guo, X.; Liu, P.; Han, J.; Ito, Y.; Hirata, A.; Fujita, T.; Chen, M. 3D Nanoporous Nitrogen-Doped Graphene with Encapsulated RuO_2 Nanoparticles for Li-O_2 Batteries. *Adv. Mater.* **2015**, *27*, 6137–6143. [CrossRef] [PubMed]
36. Su, D.; Dou, S.; Wang, G. Gold nanocrystals with variable index facets as highly effective cathode catalysts for lithium-oxygen batteries. *NPG Asia Mater.* **2015**, *7*, e155. [CrossRef]
37. Li, F.; Tang, D.M.; Tao, Z.; Liao, K.; He, P.; Golberg, D.; Yamada, A.; Zhou, H. Superior performance of a Li-O_2 battery with metallic RuO_2 hollow spheres as the carbon-free cathode. *Adv Energy Mater.* **2015**, *5*. [CrossRef]
38. Jian, Z.; Liu, P.; Li, F.; He, P.; Guo, X.; Chen, M.; Zhou, H. Core-shell-structured CNT@RuO_2 composite as a high-performance cathode catalyst for rechargeable Li-O_2 batteries. *Angew. Chem. Int. Ed. Engl.* **2014**, *53*, 442–446. [CrossRef] [PubMed]
39. Oh, S.H.; Nazar, L.F. Oxide catalysts for rechargeable high-capacity Li-O_2 batteries. *Adv. Energy Mater.* **2012**, *2*, 903–910. [CrossRef]
40. Li, F.; Wu, S.; Li, D.; Zhang, T.; He, P.; Yamada, A.; Zhou, H. The water catalysis at oxygen cathodes of lithium–oxygen cells. *Nat. Commun.* **2015**, *6*. [CrossRef]
41. Yao, K.P.C.; Risch, M.; Sayed, S.Y.; Lee, Y.-L.; Harding, J.R.; Grimaud, A.; Pour, N.; Xu, Z.; Zhou, J.; Mansour, A.; et al. Solid-state activation of Li_2O_2 oxidation kinetics and implications for Li-O_2 batteries. *Energy Environ. Sci.* **2015**, *8*, 2417–2426. [CrossRef]
42. Yao, K.P.C.; Lu, Y.-C.; Amanchukwu, C.V.; Kwabi, D.G.; Risch, M.; Zhou, J.; Grimaud, A.; Hammond, P.T.; Bardé, F.; Shao-Horn, Y. The influence of transition metal oxides on the kinetics of Li_2O_2 oxidation in Li-O_2 batteries: High activity of chromium oxides. *Phys. Chem. Chem. Phys.* **2014**, *16*, 2297–2304. [CrossRef] [PubMed]
43. Gittleson, F.S.; Sekol, R.C.; Doubek, G.; Linardi, M.; Taylor, A.D. Catalyst and electrolyte synergy in Li-O_2 batteries. *Phys. Chem. Chem. Phys.* **2014**, *16*, 3230–3237. [CrossRef] [PubMed]
44. Gittleson, F.S.; Ryu, W.-H.; Schwab, M.; Tong, X.; Taylor, A.D. Pt and Pd catalyzed oxidation of Li_2O_2 and DMSO during Li-O_2 battery charging. *Chem. Commun.* **2016**, *52*, 6605–6608. [CrossRef] [PubMed]
45. Ryu, W.H.; Gittleson, F.S.; Schwab, M.; Goh, T.; Taylor, A.D. A mesoporous catalytic membrane architecture for lithium-oxygen battery systems. *Nano Lett.* **2015**, *15*, 434–441. [CrossRef] [PubMed]
46. Schroeder, M.A.; Pearse, A.J.; Kozen, A.C.; Chen, X.; Gregorczyk, K.; Han, X.; Cao, A.; Hu, L.; Lee, S.B.; Rubloff, G.W.; et al. Investigation of the Cathode-Catalyst-Electrolyte Interface in Aprotic Li-O_2 Batteries. *Chem. Mater.* **2015**, *27*, 5305–5313. [CrossRef]
47. Huang, X.; Yu, H.; Tan, H.; Zhu, J.; Zhang, W.; Wang, C.; Zhang, J.; Wang, Y.; Lv, Y.; Zeng, Z.; et al. Carbon nanotube-encapsulated noble metal nanoparticle hybrid as a cathode material for Li-oxygen batteries. *Adv. Funct. Mater.* **2014**, *24*, 6516–6523. [CrossRef]
48. Huang, J.; Zhang, B.; Xie, Y.Y.; Lye, W.W.K.; Xu, Z.L.; Abouali, S.; Garakani, M.A.; Huang, J.Q.; Zhang, T.Y.; Huang, B.; et al. Electrospun graphitic carbon nanofibers with in-situ encapsulated Co-Ni nanoparticles as freestanding electrodes for Li-O_2 batteries. *Carbon N. Y.* **2016**, *100*, 329–336. [CrossRef]
49. Hevia, S.; Homm, P.; Cortes, A.; Núñez, V.; Contreras, C.; Vera, J.; Segura, R. Selective growth of palladium and titanium dioxide nanostructures inside carbon nanotube membranes. *Nanoscale Res. Lett.* **2012**, *7*, 342. [CrossRef] [PubMed]
50. Zhang, D.; Li, R.; Huang, T.; Yu, A. Novel composite polymer electrolyte for lithium air batteries. *J. Power Sources* **2010**, *195*, 1202–1206. [CrossRef]

51. Cecchetto, L.; Salomon, M.; Scrosati, B.; Croce, F. Study of a Li-air battery having an electrolyte solution formed by a mixture of an ether-based aprotic solvent and an ionic liquid. *J. Power Sources* **2012**, *213*, 233–238. [CrossRef]
52. Laoire, C.O.; Mukerjee, S.; Plichta, E.J.; Hendrickson, M.a.; Abraham, K.M. Rechargeable Lithium/TEGDME-LiPF-$_6$/O_2 Battery. *J. Electrochem. Soc.* **2011**, *158*, A302. [CrossRef]
53. Kichambare, P.; Kumar, J.; Rodrigues, S.; Kumar, B. Electrochemical performance of highly mesoporous nitrogen doped carbon cathode in lithium-oxygen batteries. *J. Power Sources* **2011**, *196*, 3310–3316. [CrossRef]
54. Knudsen, K.B.; Nichols, J.E.; Vegge, T.; Luntz, A.C.; McCloskey, B.D.; Hjelm, J. An Electrochemical Impedance Study of the Capacity Limitations in Na-O_2Cells. *J. Phys. Chem. C* **2016**, *120*, 10799–10805. [CrossRef]
55. Højberg, J.; McCloskey, B.D.; Hjelm, J.; Vegge, T.; Johansen, K.; Norby, P.; Luntz, A.C. An electrochemical impedance spectroscopy investigation of the overpotentials in Li-O_2 batteries. *ACS Appl. Mater. Interfaces.* **2015**, *7*, 4039–4047. [CrossRef] [PubMed]
56. Eswaran, M.; Munichandraiah, N.; Scanlon, L.G. High Capacity Li-O_2 Cell and Electrochemical Impedance. Electrochem. *Solid-State Lett.* **2010**, *13*, A121–A124. [CrossRef]
57. Bardenhagen, I.; Yezerska, O.; Augustin, M.; Fenske, D.; Wittstock, A.; Bäumer, M. In situ investigation of pore clogging during discharge of a Li/O_2battery by electrochemical impedance spectroscopy. *J. Power Sources* **2015**, *278*, 255–264. [CrossRef]
58. Yi, J.; Wu, S.; Bai, S.; Liu, Y.; Li, N.; Zhou, H. Interfacial Construction of Li_2O_2 for a Performance-Improved Polymer Li-O_2 Battery. *J. Mater. Chem. A* **2016**, *4*, 2403–2407. [CrossRef]
59. Yi, J.; Zhou, H. A Unique Hybrid Quasi-Solid-State Electrolyte for Li-O_2 Batteries with Improved Cycle Life and Safety. *ChemSusChem* **2016**, *9*, 2391–2396. [CrossRef]
60. Shui, J.-L.; Okasinski, J.S.; Kenesei, P.; Dobbs, H.A.; Zhao, D.; Almer, J.D.; Liu, D.-J. Reversibility of anodic lithium in rechargeable lithium–oxygen batteries. *Nat. Commun.* **2013**, *4*, 1–7. [CrossRef]
61. Qiao, Y.; Ye, S. Spectroscopic Investigation for Oxygen Reduction and Evolution Reactions on Carbon Electrodes in Li-O_2 Battery. *J. Phys. Chem. C* **2016**, *120*, 8033–8047. [CrossRef]
62. Thotiyl, M.M.O.; Freunberger, S.a.; Peng, Z.; Bruce, P.G. The carbon electrode in nonaqueous Li-O_2 cells. *J. Am. Chem. Soc.* **2013**, *135*, 494–500. [CrossRef] [PubMed]
63. Luo, W.B.; Pham, T.V.; Guo, H.P.; Liu, H.K.; Dou, S.X. Three-Dimensional Array of TiN@Pt3Cu Nanowires as an Efficient Porous Electrode for the Lithium-Oxygen Battery. *ACS Nano* **2017**, *11*, 1747–1754. [CrossRef] [PubMed]

Article

A New Glass-Forming Electrolyte Based on Lithium Glycerolate

Gioele Pagot [1,2], Sara Tonello [1], Keti Vezzù [1,3,4] and Vito Di Noto [1,2,3,5,*]

1 Section of Chemistry for the Technology (ChemTech), Department of Industrial Engineering, University of Padova, Via Marzolo 9, I-35131 Padova (PD), Italy; gioele.pagot@unipd.it (G.P.); sara.tonello92@gmail.com (S.T.); keti.vezzu@gmail.com (K.V.)

2 Centro Studi di Economia e Tecnica dell'Energia Giorgio Levi Cases, Via Marzolo 9, I-35131 Padova (PD), Italy

3 Consorzio Interuniversitario Nazionale per la Scienza e Tecnologia dei Materiali, Via Marzolo 1, I-35131 Padova (PD), Italy

4 Centre for Mechanics of Biological Materials—CMBM, Via Marzolo 9, I-35131 Padova (PD), Italy

5 Material Science and Engineering Department, Universidad Carlos III de Madrid, Escuela Politécnica Superior, Av.de la Universidad, 30, 28911 Leganes, Spain

* Correspondence: vito.dinoto@unipd.it; Tel.: +39-049-827-5229

Received: 19 July 2018; Accepted: 24 August 2018; Published: 1 September 2018

Abstract: The detailed study of the interplay between the physicochemical properties and the long-range charge migration mechanism of polymer electrolytes able to carry lithium ions is crucial in the development of next-generation lithium batteries. Glycerol exhibits a number of features (e.g., glass-forming behavior, low glass transition temperature, high flexibility of the backbone, and efficient coordination of lithium ions) that make it an appealing ion-conducting medium and a challenging building block in the preparation of new inorganic–organic polymer electrolytes. This work reports the preparation and the extensive investigation of a family of 11 electrolytes based on lithium glycerolate. The electrolytes have the formula $C_3H_5(OH)_{3-x}(OLi)_x$, where $0 \leq x \leq 1$. The elemental composition is evaluated by inductively coupled plasma atomic emission spectroscopy. The structure and interactions are studied by vibrational spectroscopies (FT-IR and micro-Raman). The thermal properties are gauged by modulated differential scanning calorimetry and thermogravimetric analysis. Finally, insights on the long-range charge migration mechanism and glycerol relaxation events are investigated via broadband electrical spectroscopy. Results show that in these electrolytes, glycerolate acts as a large and flexible macro-anion, bestowing to the material single-ion conductivity (1.99×10^{-4} at 30 °C and 1.55×10^{-2} S·cm^{-1} at 150 °C for x = 0.250).

Keywords: polymer electrolyte; lithium glycerolate; lithium single-ion conductor

1. Introduction

The major challenge for the improvement of lithium secondary batteries is the development of stable electrolytes capable of efficiently transferring Li ions in a wide range of temperatures. Polymer Electrolytes (PEs) seem to be the right answer thanks to their mechanical, thermal, chemical, and electrochemical stability, and good conductivity at room temperature [1,2]. Typically, with respect to ceramic [3] and liquid electrolytes [4], PEs show lower conductivity values. Nevertheless, they can be obtained as very-thin-layer systems, reducing the internal resistance of the device and thus compensating for their lower conductivity. Furthermore, the mechanical properties—in particular, their flexibility—allow the maintenance of electrode–electrolyte contact during the charge and discharge of the battery, even if the device is subjected to mechanical stress [5]. Finally, PEs and glasses have the advantage of no grain boundaries, unlike ceramics [6]. The key point to designing new and high-performing PEs is the understanding of ion dissociation phenomena, ion–ion and ion–matrix interactions, and transport events characterizing the long-range charge migration within the polymeric matrix.

Polymers that present suitable oxygen functionalities, such as polyethylene glycol (PEG) [7–10] and polyvinyl alcohol (PVA) [11], (a) present a high capability to efficiently coordinate lithium cations and (b) facilitate the ion migration events through inter- and intra-chain ion-exchange processes [2]. Glycerol (propane-1,2,3-triol) is a cheap, colorless, odorless, viscous, and nontoxic organic low-molecular-weight component endowed with (a) a glass transition (T_g) at −80 °C and (b) a melting process at 18 °C [12]. Glycerol is also thermally stable up to 195 °C [13,14] and, with its three hydroxyl groups, is able to (i) coordinate lithium ions and (ii) be easily lithiated, so giving rise to the formation of lithium alkoxide functionalities within the same molecule. Furthermore, glycerol is considered a "*glass-forming*" liquid, i.e., a material that, when it is supercooled, forms a liquid phase and shows no long-range order like normal crystalline solids. Glass-forming liquids are a category of materials that have been studied in several research areas, in particular (a) in low-temperature electrochemistry as electrolytes [15] and (b) in liquid-state physics as soft materials with peculiar molecular and ionic relaxation processes in the long-time domain [16]. These latter properties are crucial to developing a high-performing lithium ion conductor and to understanding the transport phenomena modulating the conductivity in these challenging materials.

In this work a family of 11 lithium glycerolate electrolytes with different Li contents is synthetized and extensively characterized. The general formula of these electrolytes is $C_3H_5(OH)_{3-x}(OLi)_x$, abbreviated as $GlyLi_x$ ($0 \leq x \leq 1$). Vibrational studies, carried out by FT-IR and Raman, allow for the clarification of the structure and interactions characterizing the $GlyLi_x$ electrolytes as a function of lithium concentration. The thermal stability and transitions are revealed by TGA and modulated differential scanning calorimetry (MDSC) analyses. Broadband electrical spectroscopy (BES) is used to investigate the electrical response of the electrolytes in terms of polarization and dielectric relaxations. Merging together vibrational, thermal, and BES results, a reasonable conduction mechanism is proposed for these electrolytes. Finally, the best $GlyLi_x$ electrolyte demonstrates a conductivity higher than 10^{-6} S·cm^{-1} at −10 °C, 1.99×10^{-4} S·cm^{-1} at 30 °C, and 1.55×10^{-2} S·cm^{-1} at 150 °C.

2. Results and Discussion

2.1. Chemical Composition of $GlyLi_x$ Electrolytes

During the preparation process, each glycerol molecule exchanges one H^+ with one Li^+ (see Reaction I and Section 3.1). Lithium glycerolate, C_3H_5-$(OH)_2(OLi)$, is thus obtained; this compound is indicated as GlyLi. H_2 is evolved during the reaction.

$$C_3H_5\text{-}(OH)_3 + LiH \rightarrow C_3H_5\text{-}(OH)_2(OLi) + H_2 \quad \text{(I)}$$

GlyLi is subsequently diluted with different amounts of pristine glycerol (indicated as Gly), yielding the $GlyLi_x$ electrolytes described in this work. The latter can be indicated as follows (see Equation (1)):

$$GlyLi_x = [(GlyLi)_x(Gly)_{1-x}]. \quad (1)$$

The assay of Li is obtained by ICP-AES; a summary of the composition of the 11 $GlyLi_x$ electrolytes is provided in Table 1.

Table 1. Composition of the obtained samples.

Sample	Li/wt %	n_{Li}/mol·kg^{-1}	$n_{glycerol}$/mol·kg^{-1}	$r = n_{Li}/n_o$	$x = n_{Li}/n_{glycerol}$	$y = n_{Gly}/n_{GlyLi}$
1	7.14	10.00	10.8	0.33	1.000	0
2	6.30	9.67	10.2	0.30	0.890	0.12
3	5.55	8.47	10.3	0.26	0.780	0.28
4	3.30	4.92	10.5	0.15	0.450	1.22
5	1.83	2.69	10.7	0.082	0.250	3.00
6	1.23	1.79	10.7	0.055	0.170	4.88
7	0.68	0.99	10.8	0.030	0.090	10.1
8	0.36	0.52	10.8	0.016	0.048	19.8
9	0.08	0.11	10.8	0.0033	0.010	99.0
10	0.04	0.06	10.8	0.0018	0.005	200
11	0	0	10.9	0	0	∞

2.2. Vibrational Spectroscopy Studies

2.2.1. Glycerol Conformations

In accordance with results from other studies, glycerol consists of a blend of molecules with different conformations. A brief description of the types of conformations exhibited by glycerol molecules is provided below. It is reported that three structural arrangements of CH_2OH and OH groups are possible, namely, α, β, and γ [17]. These latter derive from the allowed rotations of the CH_2OH and OH groups around the C–C covalent bond in the backbone. In the α conformation, the C–C–O angle (φ) is equal to 71°, and the terminal oxygen atom of the CH_2OH group is in *trans* position with respect to the carbon atom of the CHOH group. On the other hand, in the β conformation, the two oxygen atoms of CH_2OH and CHOH groups are in *trans* position. Finally, in the γ orientation, φ is equal to 71°, and the terminal oxygen atom of the CH_2OH group is in *trans* position with respect to the hydrogen atom of the CHOH group. The three different conformations are represented in Figure 1a.

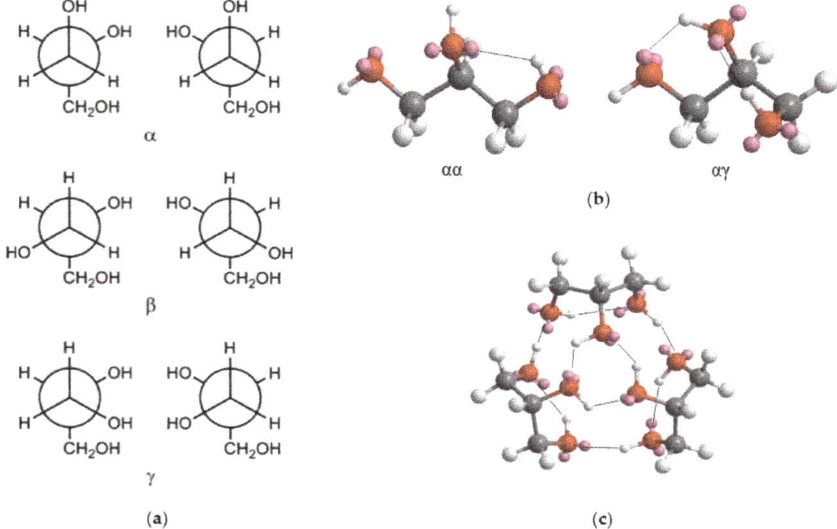

Figure 1. (**a**) The α, β, and γ conformations of glycerol; (**b**) The most probable conformations assumed by the glycerol molecule: αα and αγ. Color legend: C (grey), O (red), H (white), oxygen electron pair (pink). Hydrogen bonds are highlighted with a dotted line; (**c**) Glycerol intermolecular structure in the liquid phase.

In the crystalline form of glycerol, only the $\alpha\alpha$ conformer exists, while in the liquid both the $\alpha\alpha$ and $\alpha\gamma$ conformers are identified (Figure 1b) [18]. The $\gamma\gamma$ conformer might exist, but only with a very low probability. Moreover, dynamic mechanical studies reveal that in glycerol the intra- and inter-molecular H-bonds give rise to the formation of five-member atoms rings in the $\alpha\alpha$ and $\alpha\gamma$ conformers, while in the $\gamma\gamma$ conformer a six-atom ring coordination appears [19–21]. If we assume that (a) Li^+ cations exhibit a tetrahedral coordination geometry [22] and (b) each Li^+ is coordinated by one $-O^-$ and three electron pairs of different neutral oxygen atoms, then glycerol can assume four different coordination geometries, as follows: (i) g1: glycerol assumes an $\alpha\alpha$ conformation; one Li^+ is coordinated by four glycerol molecules through four monodentate bonds (Figure 2a); (ii) g2.1: glycerol is characterized by an $\alpha\alpha$ conformation; one glycerol molecule coordinates Li^+ as a bidentate ligand and two more glycerol molecules coordinate Li^+ as monodentate ligands (Figure 2b); (iii) g2.2: glycerol exhibits an $\alpha\alpha$ conformation; two glycerol molecules coordinate Li^+ as bidentate ligands (Figure 2c); and (iv) g3: glycerol has a $\gamma\gamma$ conformation; one glycerol molecule coordinates Li^+ as a monodentate ligand, while another glycerol molecule coordinates Li^+ as a tridentate ligand (Figure 2d). All of these coordination geometries are present in the electrolytes described in this work. FT-IR and micro-Raman studies allow us to understand which coordination geometry is predominant at each n_{Li}/n_{gly} ratio (x).

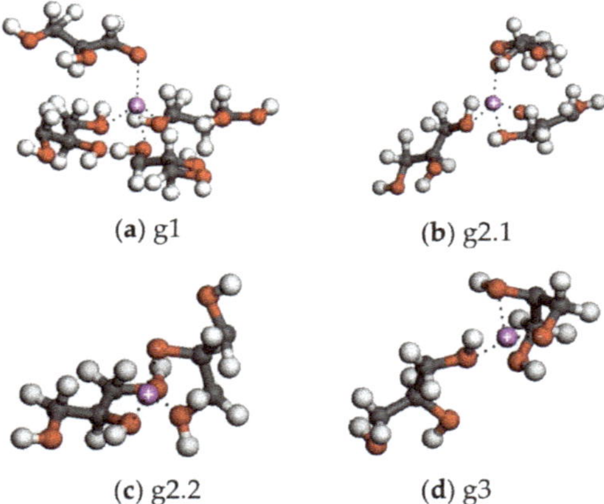

Figure 2. Coordination geometries assumed by glycerol molecules upon coordination of Li^+ cations. Color legend: C (grey), O (red), H (white), Li (violet). Li–O interactions are highlighted with a dotted line. (**a**) g1 coordination geometry, with four monodentate glycerol ligands; (**b**) g2.1 coordination geometry, with two monodentate glycerol ligands and one bidentate glycerol ligand; (**c**) g2.2 coordination geometry, with two bidentate glycerol ligands; (**d**) g3 coordination geometry, with one monodentate glycerol ligand and one tridentate glycerol ligand.

2.2.2. Fourier Transform Infrared Spectroscopy

Vibrational spectroscopies are a useful tool to identify which of the coordination geometries represented in Figure 2 predominates at each concentration of Li^+ in the different electrolytes. FT-IR spectra are presented in Figure 3.

The vibrational spectra can be divided into three different regions: (i) at wavenumbers higher than 3000 cm^{-1}, the typical OH stretching vibrations (ν_{OH}) are observed [19,23,24]; (ii) in the wavenumber range 2700–3000 cm^{-1}, the CH stretching vibrations (ν_{CH}) of the glycerol backbone are revealed [19,23,24]; (iii) finally, between 480 and 1500 cm^{-1}, the CH_2 scissoring (sr), OH in-plane bending (δ_{ip}), CO stretching (ν_{CO}), CO bending (δ_{CO}), CC internal rotations (φ_{CC}), and CO rocking (ϱ_{CO}) vibrations are detected [19,23–26]. Table 2 reports a complete correlative assignment of all the vibrational modes shown in Figure 3.

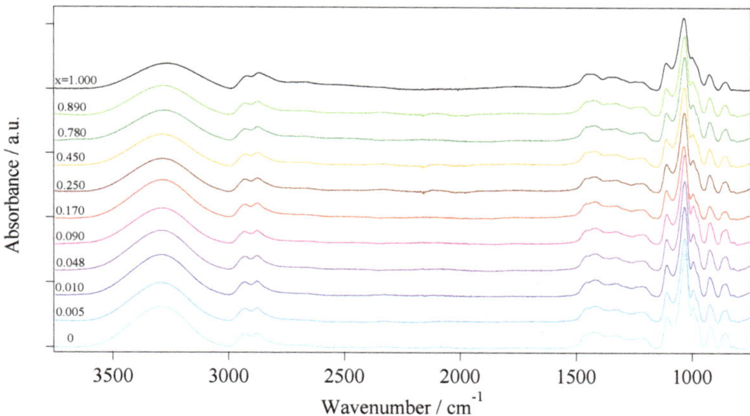

Figure 3. FT-IR spectra of $GlyLi_x$ electrolytes.

Table 2. FT-IR and Raman correlational assignments of glycerol (Gly) and lithium glycerolate (GlyLi) samples.

Glycerol, Gly (x = 0)		Lithium Glycerolate, GlyLi (x = 1.000)			
ω_{IR} [1]/cm^{-1}	ω_{Raman}/cm^{-1}	ω_{IR}/cm^{-1}	ω_{Raman}/cm^{-1}	Assignment [2]	Ref.
3422 (s)	–	3386 (s)	–	ν^{f}(OH)	[19,23,24]
3286 (vs)	3350 (s)	3267 (vs)	3350 (s)	ν^{Hy}(OH)	[19,23,24]
–	–	3149 (m)	–	ν^{g3Li}(OH)	[this work]
–	–	3071 (w)	–	ν^{g2Li}(OH)	[this work]
–	2966 (wv)	–	2967 (w)	$\nu^{g2,a}$(CH)	[this work]
2931 (m)	2945 (vs)	2929 (m)	2945 (vs)	$\nu^{g1,a}$(CH)	[19,23,24]
–	2909 (w)	–	2909 (s)	$\nu^{g2,s}$(CH)	[this work]
–	2885 (vs)	–	2885 (vs)	$\nu^{g1,s}$(CH)	[19,23,24]
2879 (m)	2874 (m)	2878 (m)	2876 (m)	$\nu^{g3,a}$(CH)	[19,23,24]
–	–	2867 (m)	–	$\nu^{g3,a}$(CH)	[this work]
2844 (m)	2831 (s)	2842 (w)	–	$\nu^{g3,s}$(CH)	[this work]
2800 (w)	–	–	–	$\nu^{g3,s}$(CH)	[this work]
2754 (vw)	2741 (vw)	2695 (vw)	2736 (w)	ν^{s}(CH of C_2)	[19,23,24]
2727 (vw)	–	–	–	ν^{s}(CH of C_2)	[this work]
–	–	2504 (vw)	–	ν(CH)	[this work]
–	1465 (s)	–	1464 (s)	sr(CH_2)	[19,23,24]
1416 (s)	–	1435 (s)	–	δ_{ip}(OH),ω(CH_2)	[19,23,24]
1323 (m)	1313 (vw)	1336 (m)	1313 (vw)	δ_{ip}(OH),ω(CH)	[19,23,24]
1210 (w)	1253 (v)	1225 (vs)	1253 (v)	δ_{ip}(OH),ω(CH)	[19,23,24]
1108 (vs)	1112 (s)	1111 (m)	1112 (s)	$\nu^{g3,sec}$(C–C–O)	[19,23,24]
1088 (w)	–	1091 (w)	–	$\nu^{g1,sec}$(C–C–O)	[this work]
1075 (w)	–	1073 (w)	–	$\nu^{g2,sec}$(C–C–O)	[this work]
1054 (m)	–	1054 (m)	–	$\nu^{g3,pri}$(C–C–O)	[this work]
1029 (vs)	1055 (s)	1034 (vs)	1055 (s)	$\nu^{g1,pri}$(C–C–O)	[19,23,24]
995 (s)	975 (w)	998 (sh)	975 (w)	ν^{a}(CO),T	[19,23,24]
977 (s)	–	980 (s)	–	ν^{s}(CO),T	[this work]
924 (m)	922 (m)	925 (m)	922 (m)	ν^{a}(CO),G	[19,23,24]
908 (vw)	–	909 (vw)	–	ν^{s}(CO),G	[this work]
866 (w)	–	867 (w)	–	$\nu^{a,pri}$(C–C–C)	[this work]
849 (m)	850 (s)	850 (m)	850 (s)	$\nu^{s,pri}$(C–C–C)	[19,23,24]
819 (vw)	820 (m)	820 (vw)	820 (m)	ν^{pri}(C–C–C)	[19,23,24]
639 (m, sh)	673 (w)	630 (m, sh)	673 (w)	δ(C–C–O)	[19,23,24]
550 (s)	548 (w)	563 (w)	548 (w)	φ(C–C)	[19,25]
488 (s)	484 (m)	493 (m)	484 (m)	ϱ(C–C–O)	[19,25]
	413 (m)		413 (m)	ϱ(C–C–O)	[19,25]
	327 (vw)		327 (vw)	φ(C–C), δ(C–C–O)	[19]

[1] Relative intensities are shown in parentheses: vs: very strong; s: strong; m: medium; w: weak; vw: very weak; sh: shoulder. [2] ν: stretching; δ: bending; sr: scissoring; ω: wagging; φ: internal rotations; ϱ: rocking a: antisymmetric mode; s: symmetric mode; f: free; Hy: involved in a hydrogen bonding; Li: coordinate to lithium; g1: g1 coordination geometry; g2: g2 coordination geometry; g3: g3 coordination geometry; ip: in-plane; pri: primary alcohol; sec: secondary alcohol; T: *trans*; G: *gauche*.

The OH stretching vibration centered at ca. 3300 cm^{-1} and the CO stretching vibrations of primary and secondary alcohols peaking in the range 1000–1100 cm^{-1} are the most affected by the different concentration of Li^+. This is even more evident from the differential spectra reported in Figure S1 of Supplementary Materials, where the spectrum of pristine glycerol is subtracted from the spectrum of each sample to highlight the differences between Gly and the $GlyLi_x$ electrolytes. It is observed that the ν(OH) vibration shifts to lower wavenumbers as the concentration of Li^+ is raised; the intensity of the ν(OH) vibration is concurrently reduced. This trend is easily explained considering that the OH stretching vibrations are involved in the formation of hydrogen bonds, whose number decreases upon the addition of Li^+ [27]. Furthermore, the downshift in wavenumber of ν(OH) is the result of a weakening of the hydrogen–oxygen bond strength in glycerol [27]. The second FT-IR peaks that are strongly affected by the concentration of Li^+ are those centered at ca. 1034 and 1111 cm^{-1}, which are attributed to the CO stretching of the g1 and g3 conformation geometries, respectively. It is observed that the intensities of both vibrations decrease as the Li^+ content is raised. In particular, the latter decreases with a lower magnitude with respect to the former, indicating that, probably, the increase in the lithium content results in the stabilization of the g3 conformation geometry. These trends are plotted in Figure S2 of Supplementary Materials. In summary, the presence of Li^+ destabilizes the typical conformation geometry of pristine glycerol, modifying the hydrogen–oxygen interactions.

Further insight into the glycerol conformation geometries at different concentrations of Li^+ can be obtained by the Gaussian decomposition of the peaks in the 2500–3700 and 770–1150 cm^{-1} regions. For the sake of brevity, only the results obtained from pristine glycerol, Gly (x = 0), and lithium glycerolate, GlyLi (x = 1.000), are reported in Figure 4.

With respect to the spectrum of pristine glycerol (Figure 4a), the spectra of both GlyLi (Figure 4b) and $GlyLi_x$ electrolytes with $x \geq 0.170$ reveal two new peaks at ca. 3071 and 3149 cm^{-1}. These latter peaks are attributed to the OH stretching vibrations of the g2 (Figure 2b,c) and g3 (Figure 2d) coordination geometries of the glycerol molecules, respectively. On the basis of the fitting results, a semiquantitative analysis is carried out based on the areas of the peaks centered at ca. 3386, 3267, 3149, and 3071 cm^{-1}. The areal percentages are calculated following Equation (2):

$$A_n\% = \frac{A_n}{\sum_{i=1}^{m} A_i} \cdot 100, \tag{2}$$

where $A_n\%$ is the areal percentage of the peak n, A_n is the calculated area of the peak n, and the sum takes into consideration the areas of all the peaks centered at ca. 3386, 3267, 3149, and 3071 cm^{-1}. Results are reported in Figure 5a.

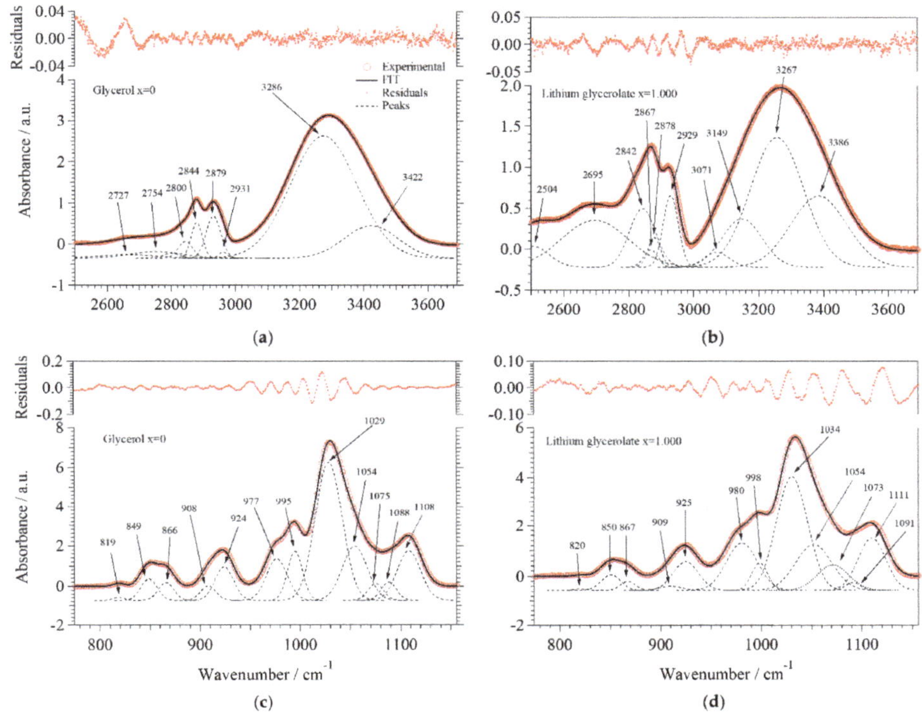

Figure 4. Gaussian fitting of the FT-IR spectra of (**a**) pristine glycerol, Gly (x = 0), in the 2500–3700 cm^{-1} region; (**b**) lithium glycerolate, GlyLi (x = 1.000), in the 2500–3700 cm^{-1} region; (**c**) pure glycerol, Gly (x = 0), in the 770–1150 cm^{-1} region; and (**d**) lithium glycerolate, GlyLi (x = 1.000), in the 770–1150 cm^{-1} region.

On the basis of these results, the proposed electrolytes can be divided into three groups, depending on the concentration of Li^+:

- Group A: At low concentrations of Li^+ ($0 \leq x \leq 0.090$), the areal percentage of the peaks associated to ν^f(OH) (3386 cm^{-1}) and ν^{Hy}(OH) (3267 cm^{-1}) varies by no more than 5%. In this group of electrolytes, A_{3149} and A_{3071}, which are attributed respectively to the ν^{g3Li}(OH) and ν^{g2Li}(OH) modes, are equal to 0. This suggests that at low values of x, a large number of intra- and inter-molecular hydrogen bonds are present. The coordination geometry remains similar to that of pristine glycerol, i.e., g1 (Figure 2a). Indeed, g1 allows for a higher number of hydrogen bonds with respect to g2 and g3 coordination geometries.
- Group B: In the middle of the concentration range of Li^+ ($0.170 \leq x \leq 0.450$), A_{3386} remains almost constant while A_{3267} is decreased by ca. 63%, indicating a lower number of hydrogen bonds. A_{3149} and A_{3071} appear in the spectra of these samples, reaching values of ca. 30%. The decrease in the number of hydrogen bonds is attributed to the increased number of oxygen atoms involved in the coordination of Li^+, as confirmed by the increased intensities of the peaks attributed to the ν^{g3Li}(OH) and ν^{g2Li}(OH). In samples with $0.170 \leq x \leq 0.450$, the predominant coordination geometries are g2.1 and g2.2 (Figure 2c,d) due to a lower number of hydrogen bonds and to the formation of lithium coordination by oxygen functionalities.
- Group C: In the high Li^+ concentration range ($0.780 \leq x \leq 1.000$), A_{3386} remains almost constant, and A_{3267} goes back to medium values of areal percentage (up to 64%). Moreover, in this region the areas of ν^{g3Li}(OH) and ν^{g2Li}(OH) peaks begin to become clearly different, with the former

larger than the latter. This evidence is easily explained if we consider that at x ≥ 0.780 the prevailing coordination geometry is g3. This allows for a higher number of hydrogen bonds with respect to g2; at the same time, a significant number of oxygen atoms is coordinating Li^+.

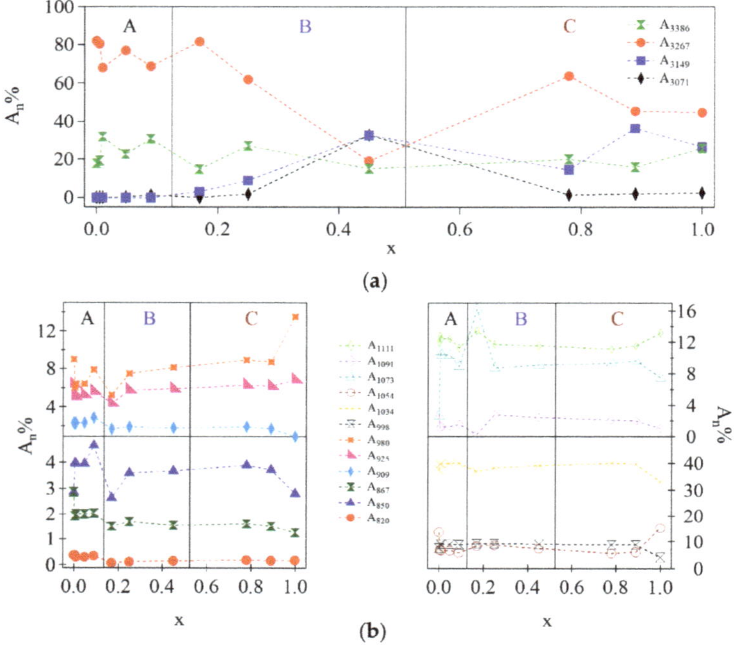

Figure 5. (**a**) Semiquantitative analysis of the peaks centered at ca. 3386, 3267, 3149, and 3071 cm^{-1} as a function of x; (**b**) Semiquantitative analysis of the peaks in the range 770–1150 cm^{-1} as a function of x.

Equation (2) is used also to evaluate the behavior of peak areas at different Li^+ concentrations in the range 770–1150 cm^{-1}; the results are reported in Figure 5b. The most interesting peaks to study the different coordination geometries assumed by the glycerol molecules are those centered at ca. 1034, 1054, 1073, 1091, and 1111 cm^{-1}, attributed to $\nu^{g1,pri}$(C–C–O), $\nu^{g3,pri}$(C–C–O), $\nu^{g2,sec}$(C–C–O), $\nu^{g1,sec}$(C–C–O), and $\nu^{g3,sec}$(C–C–O), respectively.

- Group A: At low concentrations of Li^+ ($0 \leq x \leq 0.090$), the $\nu^{g1,pri}$(C–C–O) vibration dominates the spectra, indicating that the majority of Li^+ is coordinated by the oxygen functionalities of primary alcohols in the g1 coordination geometry.
- Group B: In the middle of the concentration range of Li^+ ($0.170 \leq x \leq 0.450$), we observe a decrease in the area of the peak attributed to $\nu^{g1,pri}$(C–C–O) vibration, concurrently with a significant increase of both A_{1073} (here the effect is significant) and A_{1111} (here the effect is visible, but less pronounced). Indeed, in this region the main coordination geometries assumed by glycerol molecules are g2.1 and g2.2, with secondary alcohol hydroxyl groups coordinating Li^+.
- Group C: Finally, at a high concentration of Li^+ ($0.780 \leq x \leq 1.000$), the areas of the peaks attributed to $\nu^{g1,pri}$(C–C–O), $\nu^{g3,pri}$(C–C–O), and $\nu^{g3,sec}$(C–C–O) increase. On the contrary, the areas of the peaks associated to $\nu^{g2,sec}$(C–C–O) and $\nu^{g1,sec}$(C–C–O) show a minimum for x = 1.000. This demonstrates that at a high concentration of Li^+ the dominating coordination geometry for glycerol molecules is g3.

The results obtained from the fitting of the 2500–3700 cm^{-1} region are in accordance with those obtained from the analysis of the spectra in the 770–1150 cm^{-1} range. Table 3 summarizes the different coordination geometries assumed by the glycerol molecules as a function of x.

Table 3. Coordination geometries and thermal results of Gly and GlyLi$_x$ electrolytes.

x	Group	Coordination Geometry [1]	$T_{dec.}$ [2]/°C	T_g [3]/°C	ΔH_{er} [4]/J·g^{-1}
0	A	g1	195.09	−78.51	−2.476
0.005	A	g1	195.03	−78.22	−2.267
0.010	A	g1	190.38	−78.53	−2.521
0.048	A	g1	190.19	−76.18	−1.877
0.090	A	g1	188.95	−76.17	−2.057
0.170	B	g2.1 and g.2.2	172.77	−72.35	−2.096
0.250	B	g2.1 and g.2.2	174.79	−69.07	−2.077
0.450	B	g2.1 and g.2.2	170.14	−66.15	−1.84
0.780	C	g3	191.34	−64.93	−1.637
0.890	C	g3	177.09	−63.04	−3.483
1.000	C	g3	185.42	−42.95	−3.29

[1] Results obtained from the fitting of FT-IR and micro-Raman spectra. [2] Decomposition temperature obtained from TGA measurements. [3] Glass transition temperatures are evaluated from modulated differential scanning calorimetry (MDSC) measurements. [4] Relaxation enthalpies are calculated from MDSC measurements.

2.2.3. Micro-Raman Spectroscopy

With respect to FT-IR, micro-Raman spectroscopy measurements are more sensitive to CH vibrational modes. Thus, the decomposition of the micro-Raman spectra reported in Figure 6 yields additional information on the coordination geometries taken by glycerol as a function of the concentration of Li$^+$.

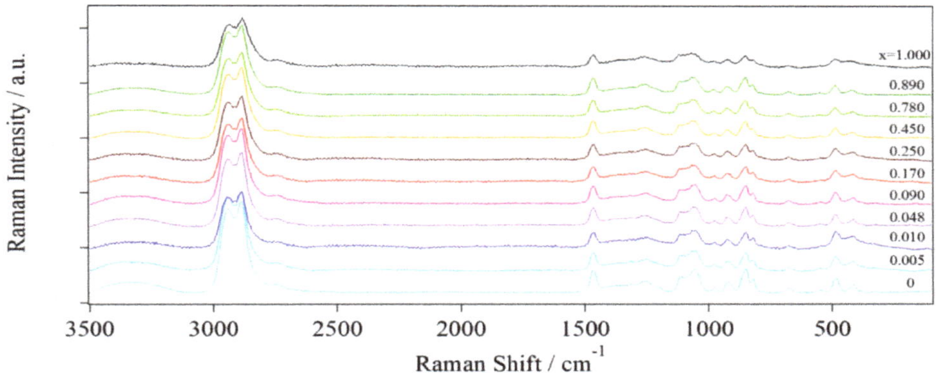

Figure 6. Normalized micro-Raman spectra of GlyLi$_x$ samples.

The Lorentzian fitting of the spectra of pristine glycerol in the 2600–3100 cm^{-1} range and the areal behavior of the peaks as a function of x are reported in Figures S3 and S4 of the Supplementary Materials. In accordance with FT-IR results, in micro-Raman spectra the electrolytes are clearly divided into three different groups, depending on the concentration of Li$^+$:

- Group A: At low concentrations of Li$^+$ ($0 \leq x \leq 0.090$), A_{2945}, attributed to the $\nu^{g1,a}$(CH) vibration, has the highest value among all the other peaks. Again, it is confirmed that at these concentrations of Li$^+$ the predominant coordination geometry taken by glycerol molecules is g1.
- Group B: In the middle of the concentration range of Li$^+$ ($0.170 \leq x \leq 0.450$), it appears that the area attributed to the CH stretching of g1 conformation decreases, while the highest increase is

observed for A_{2966} and A_{2909}. The latter are attributed to $\nu^{g2,a}$(CH) and $\nu^{g2,s}$(CH), respectively, with a maximum at x = 0.250. In this concentration region, glycerol molecules prefer g2.1 and g2.2 coordination geometries.

- Group C: Finally, at a high concentration of Li^+ (0.780 ≤ x ≤ 1.000), a further decrease of the area attributed to $\nu^{g1,a}$(CH) vibration is observed. Differently from Group B, a significant increase of A_{2831} is observed, attributed to $\nu^{g3,s}$(CH). This indicates that in this concentration range of Li^+, the number of glycerol molecules taking g3 coordination geometry is increased.

In summary, FT-IR analysis is useful to studying how OH and CO stretching vibrations are affected by the different concentrations of Li^+. On the other hand, micro-Raman spectroscopy allows us to evaluate the changes in coordination geometry, studying the CH stretching modifications. A clear division into three groups of samples with a different concentrations of Li^+ observed: at low Li^+ content (x ≤ 0.090), glycerol molecules prefer the g1 coordination geometry (Figure 2a), while at medium concentrations of Li^+ (0.170 ≤ x ≤ 0.450), the prevalent coordination geometry is g2 (Figure 2b,c). Finally, at high lithium content (x ≥ 0.780), g3 (Figure 2d) is the most probable conformation for glycerol molecules.

2.3. Thermal Studies

2.3.1. Thermo-Gravimetric Analysis (TGA)

Thermo-gravimetric analyses allow for the determination of the thermal stability of the proposed $GlyLi_x$ electrolytes, and for evaluation of the effects of the presence of Li^+ on the decomposition temperature. Results are reported in Figure 7a and Table 3.

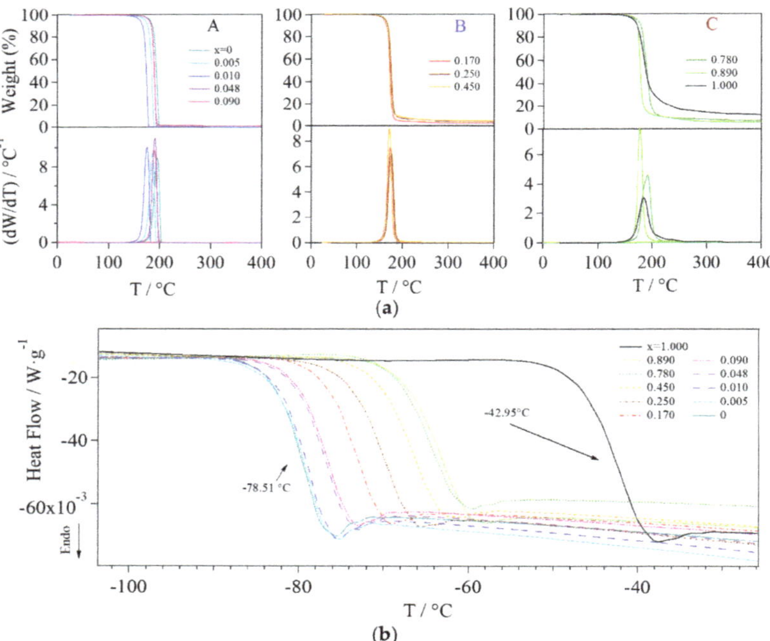

Figure 7. (**a**) Thermo-gravimetric analyses of $GlyLi_x$. (**b**) MDSC profiles of $GlyLi_x$.

In accordance with the literature, pristine glycerol shows the highest thermal stability among all the samples, with a decomposition temperature ($T_{dec.}$) of 195.09 °C and no residual weight [28].

When lithium is added, a high-temperature residue appears, the magnitude of which increases as the lithium content is raised. This evidence is interpreted as admitting that the high-temperature residue consists of lithium oxide, which is formed at high temperatures. Li_2O can be formed under nitrogen atmosphere due to the decomposition of the glycerol matrix. The trend of the decomposition temperatures as a function of x reflects the division into the three groups of electrolytes also revealed by the vibrational spectroscopies (see Table 3). In Group A, all the samples are characterized by a decomposition temperature that is similar to that of pristine glycerol. When x reaches 0.170, a drop of more than 16 °C in the thermal stability is observed. In Group C, $T_{dec.}$ increases again to high values. On the basis of these results, it is deduced that (i) the introduction of Li^+ lowers the thermal stability of the electrolytes, promoting the decomposition of glycerol, and (ii) the thermal stability of the electrolytes is correlated to the coordination geometry assumed by the glycerol molecules. Indeed, a decrease in the decomposition temperature of $GlyLi_x$ could be ascribed to the reduction of the number of hydrogen bonds [29].

2.3.2. Modulated Differential Scanning Calorimetry (MDSC) Studies

Figure 7b shows the MDSC results of the synthetized materials. In all the samples, the presence of two different contributions is observed; these are easily decoupled thanks to the modulated option of the DSC instrument. The separation of reversible and nonreversible heat flow components for pristine glycerol (x = 0) and GlyLi (x = 1.000) are reported in Figure S5 of the Supplementary Materials. The reversible contribution is attributed to the glass transition (T_g) of the glycerol matrix, while the nonreversible component is due to an enthalpic relaxation (ΔH_{er}) associated to T_g. MDSC measurements reveal that with increasing lithium content (x), the T_g of the materials increases (see Table 3), with a difference of more than 35 °C between pristine glycerol ($T_g = -78.51$ °C, similar to that reported in the literature [30]) and GlyLi ($T_g = -42.95$ °C). This indicates that Li^+, coordinated by glycerol molecules, is able to create a coordination network that increases the energy needed for the structural reorganization of the electrolyte at T_g. The enthalpic relaxation value calculated for pristine glycerol, $-2.476\ J\cdot g^{-1}$, is in accordance with the literature [31]. All the electrolytes exhibit a negative value of ΔH_{er} (see Table 3), indicating that the transition is endothermic. It is also observed that the coordination geometries assumed by the glycerol molecules influence the enthalpy associated with the glass transition. In Group A, the high number of hydrogen bonds keeps the enthalpy low. As the content of Li^+ is increased (Group B), the number of hydrogen bonds is reduced. Correspondingly, ΔH_{er} values are raised. Finally, in Group C, the negative trend of ΔH_{er} as x is raised is explained by the formation of several tetrahedral coordination geometries due to the presence of a high concentration of Li^+.

2.4. Broadband Electrical Spectroscopy

The electric response of the electrolytes in terms of polarization and relaxation phenomena is investigated by BES. These measurements are useful to studying the interplay between the structure, relaxations, and conduction mechanism of the proposed electrolytes as a function of the concentration of Li^+. Measurements are carried out between −130 and 150 °C, in the 10^{-1}–10^7 Hz frequency range. The spectra of the complex permittivity versus temperature and frequency are fitted with the following empirical equation [32]:

$$\varepsilon^*(\omega) = i\left(\frac{\sigma_0}{\varepsilon_0\omega}\right)^N + \sum_{k=1}^{n}\frac{\sigma_k(i\omega\tau_k)^{\gamma_k}}{i\omega[1+(i\omega\tau_k)^{\gamma_k}]} + \sum_{j=1}^{m}\frac{\Delta\varepsilon_j}{i\omega\left[1+(i\omega\tau_j)^{\alpha_j}\right]^{\beta_j}} + \varepsilon_\infty \tag{3}$$

where $\varepsilon^*(\omega) = \varepsilon'(\omega) - i\varepsilon''(\omega)$ ($\sigma^*(\omega) = i\omega\varepsilon^*(\omega)$). The first term of Equation (3) is attributed to the conductivity of the electrolyte at zero frequency (this is the residual conductivity of the sample). ε_∞ accounts for the permittivity of the electrolyte at infinite frequency (electronic contribution). The second term is associated with the electrode (k = 1) and interdomain (k $\geq$ 2) polarization phenomena. τ_k and σ_k are the relaxation times and the conductivity, respectively, associated with the *k*th polarization.

γ_k is a pre-exponential factor that varies from 0.5 to 1.0. The third term, attributed to the dielectric relaxation, is described by the Havriliak–Negami theory [32,33]. $\Delta\varepsilon$, τ_j, α_j, and β_j correspond to the dielectric strength, relaxation time, and symmetric and antisymmetric shape parameters of the jth relaxation event, respectively.

The 3D tanδ (tanδ = $\varepsilon''/\varepsilon'$) profiles are adopted to identify the polarization and relaxation phenomena occurring in the electrolytes. The 3D tanδ profiles of four samples—$GlyLi_x$ with x = 0, 0.010, 0.170, and 1.000, belonging to the three groups of electrolytes A, B, and C—are reported in Figure 8. The 3D tanδ profiles of the other electrolytes are found in Figure S6 of Supplementary Materials.

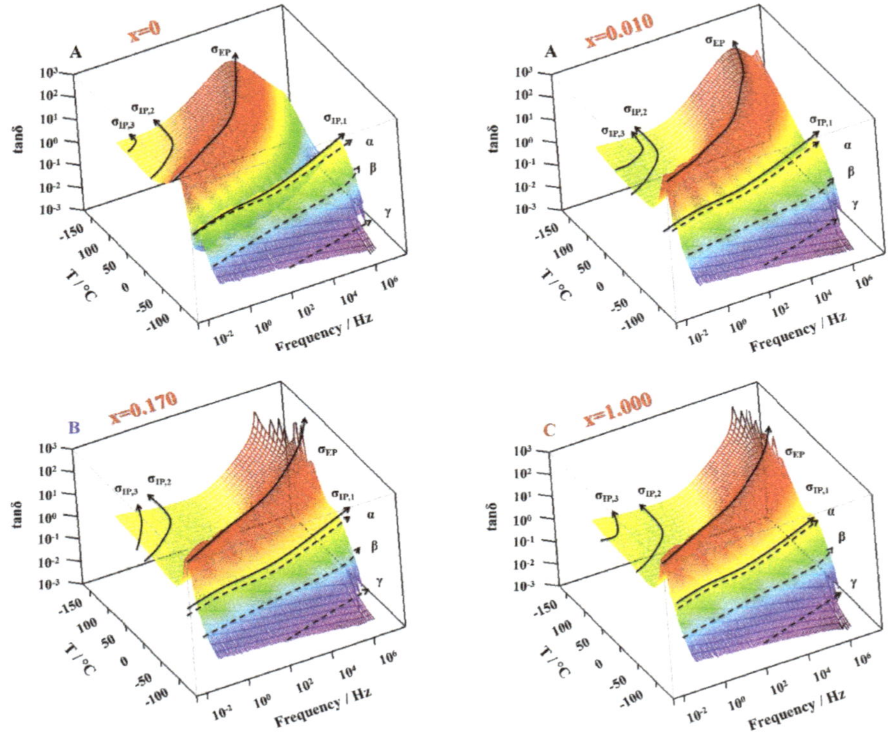

Figure 8. 3D tanδ surfaces of $GlyLi_x$ samples with x = 0, 0.010, 0.170, and 1.000.

2.4.1. Polarization Phenomena

Four different polarization phenomena are observed: (i) the electrode polarization event (σ_{EP}), which is observed at temperatures higher than −60 °C and corresponds to the accumulation of charge at the interface between the sample and the platinum blocking electrodes; and (ii) three different interdomain polarizations at temperatures ranging between −70 and 10 °C ($\sigma_{IP,1}$), between 0 and 150 °C ($\sigma_{IP,2}$), and above 60 °C ($\sigma_{IP,3}$). These polarizations are attributed to the accumulation of charge at the interface between domains with different permittivity. The presence of these latter polarization phenomena (σ_{IP}) indicates that the electrolytes are heterogeneous at the mesoscale and glycerol molecules tend to aggregate, forming nano-domains.

The conductivity values of each polarization, obtained from the fitting parameter associated to all the observed events, are plotted versus T^{-1}. Each curve is fitted with a suitable model, i.e., Vogel–Tammann–Fulcher (VTF) [34] or Arrhenius (A) equation [32]. Results are shown in Figure 9.

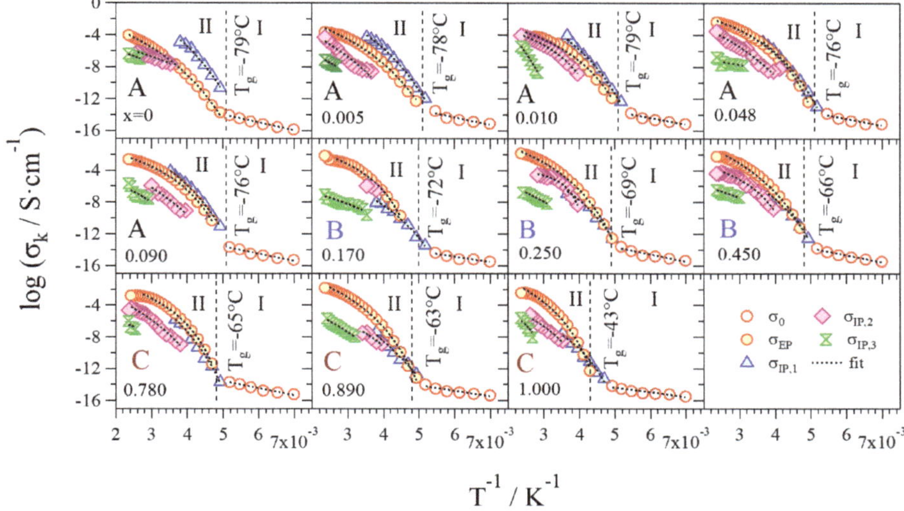

Figure 9. Log σ_k vs T^{-1} curves of each polarization for every sample. Regions I and II are divided by the T_g of the sample.

At temperatures lower than the T_g, i.e. ca. −60 °C (Region I), the only contribution to the overall conductivity is provided by σ_0 and follows an Arrhenius behavior. This evidence witnesses that in this temperature region, charge migration occurs through hopping events between different domains. At temperatures higher than T_g (Region II), all σ_k reveal a VTF-like behavior, indicating that the conduction mechanism is assisted by the segmental motion of the cluster aggregates. The overall conduction is the sum of the single conductivities of each polarization occurring in the electrolyte (see Equation (4)) [8]:

$$\sigma_T = \sigma_0 + \sum_{k=1}^{n} \sigma_k = \sigma_0 + \sigma_{EP} + \sigma_{IP,1} + \sigma_{IP,2} + \sigma_{IP,3}. \quad (4)$$

Figure 10 shows the logarithm of the overall conductivity as a function of T^{-1}; the $GlyLi_x$ samples are divided into the three groups described in Section 2.2.2 on the basis of the concentration of Li^+.

- Group A: Pristine glycerol and electrolytes with a low content of Li^+ ($x \leq 0.090$) show the lowest values of overall conductivity. At low concentrations of Li^+, glycerol molecules are in the $\alpha\alpha$ conformation and coordinate Li^+ through the g1 geometry, where the highest number of intra- and inter-molecular hydrogen bonds is achieved. Thus, the presence of these interactions gives rise to glycerol structures that hinder the migration of Li^+. Furthermore, a low concentration of charge carriers decreases the overall conductivity. In Group A, a conductivity drop is observed at ca. 0 °C, even if MDSC measurements do not reveal any thermal transition. This phenomenon can be explained by considering that, at these concentrations of Li^+ and in this particular temperature range ($T_g \leq T \leq 0$ °C), the structure assumed by the glycerol molecules facilitates the $\sigma_{IP,1}$ polarization, making it the component providing the highest contribution to the overall conductivity (Figure 9) and perturbing the log σ_T vs T^{-1} curve behavior. This effect decreases as the content of Li^+ is increased. At temperatures higher than 0 °C, the main contribution to the overall conductivity is mostly provided by the electrode polarization.
- Group B: In this concentration region the highest values of overall conductivity are reached. In particular, $GlyLi_{0.250}$ shows the highest lithium single-ion room temperature conductivity value, i.e., 1.99×10^{-4} at 30 °C. This value is in line with those present in the literature for

lithium-ion-conducting polymer electrolytes. This makes the proposed electrolytes extremely interesting for application in Li-ion batteries [19,35]. Furthermore, this electrolyte is characterized by (i) an impressive high-temperature conductivity of 1.55×10^{-2} $S{\cdot}cm^{-1}$ at 150 °C; and (ii) a significant low-temperature conductivity, that is higher than 10^{-6} $S{\cdot}cm^{-1}$ at −10 °C, thanks to the glass-forming behavior of the glycerol matrix. In this concentration range of Li^+, the g2 conformation geometry assumed by glycerol molecules does not facilitate the $\sigma_{IP,1}$ polarization. Indeed, (i) the conductivity of this event strongly decreases; (ii) σ_{EP} becomes the most contributing component to the overall conductivity at $T > T_g$; and (iii) the conductivity drop at 0 °C is no longer detected.

- Group C: Similar to Group B, the electrolytes belonging to this concentration region demonstrate high values of room temperature conductivity ($\sigma_T > 10^{-4}$ $S{\cdot}cm^{-1}$). Nevertheless, GlyLi shows a decrease in conductivity, especially in the $T_g \leq T \leq 30$ °C temperature range. This is probably the result of three concurring effects: (i) as already demonstrated by MDSC measurements, the introduction of a high number of Li^+ ions makes the structure of the glycerol host more rigid, hindering the migration process of Li^+; (ii) the g3 conformation geometry, which is assumed by glycerol molecules in this concentration region, strongly decreases the $\sigma_{IP,1}$ polarization contribution and thus lowers the overall conductivity; and (iii) the presence of a huge number of Li^+ ions increases the probability to entrap lithium ions into the tetrahedral coordination structures of the glycerol molecules.

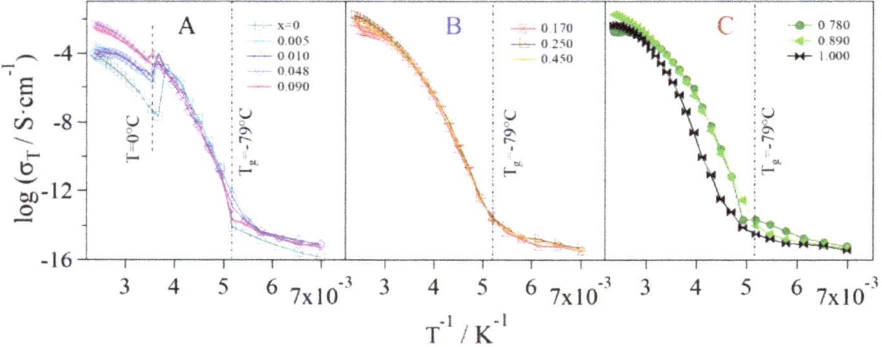

Figure 10. Log σ_T vs T^{-1} curves for the various groups of $GlyLi_x$ samples. Group A: $x \leq 0.090$; Group B: $0.170 \leq x \leq 0.450$; Group C: $0.780 \leq x \leq 1.000$.

In summary, at temperatures lower than the T_g (i.e., ca. −60 °C), the only contribution to the overall conductivity is provided by σ_0, and Li^+ are exchanged with a hopping process. At high temperatures (T > 10 °C), σ_{EP} mainly contributes to the overall conductivity and the migration of Li^+ is assisted by the segmental motions of the glycerol aggregates. Finally, at medium temperatures ($T_g \leq T \leq 10$ °C) and at a low concentration of Li^+ ($x \leq 0.090$), the main contribution to the overall conductivity is given by $\sigma_{IP,1}$. The $\sigma_{IP,1}$ contribution decreases as the content of Li^+ is raised, while σ_{EP} becomes more and more intense, thus providing the largest contribution to the overall conductivity.

2.4.2. Relaxation Events

At temperatures lower than 10 °C, three different dielectric relaxation events are observed, belonging to glycerol and $GlyLi_x$ samples (Figure 8 and Figure S6): the relaxation that appears at high frequency (>10^2 Hz) and at temperatures lower than −70 °C is attributed to a γ dielectric relaxation, and it is associated to the rapid local fluctuations of the dipole moments of glycerol hydroxyl terminal functionalities (Figure S7 of the Supplementary Materials) [36]; at lower frequencies with respect to γ, and in the temperature range between −100 and −20 °C, one β dielectric relaxation is observed. This relaxation is attributed to an internal modification in the molecular conformation of glycerol, and it is typically referred to as a "*β mode of glass-forming materials*" [37,38]. In particular, it is attributed to the local fluctuations of the dipole moments of glycerol molecules that are coordinating protons or lithium cations through the hydrogen bonds (see Figure S7 of the Supplementary Materials) [8]. Finally, the relaxation that is observed at the lowest values of the frequency and at temperatures up to 30 °C is assigned to an α dielectric relaxation mode. This type of relaxation is a dynamic process involving the diffusion of conformational states along the polymer chains and aggregates, and it is known as a "*segmental motion*" (Figure S7) [37].

The dielectric relaxation frequency (f_j) and dielectric strength ($\Delta\varepsilon_j$) values, obtained from the fitting of the 3D surfaces of Figure 8 and Figure S6 using Equation (3), are plotted as a function of temperature (T^{-1}) and reported in Figure 11 and Figure S8 of the Supplementary Materials, respectively. All the curves present in Figure 11 are fitted with Vogel–Tammann–Fulcher–Hesse (VTFH) or Arrhenius-like equations.

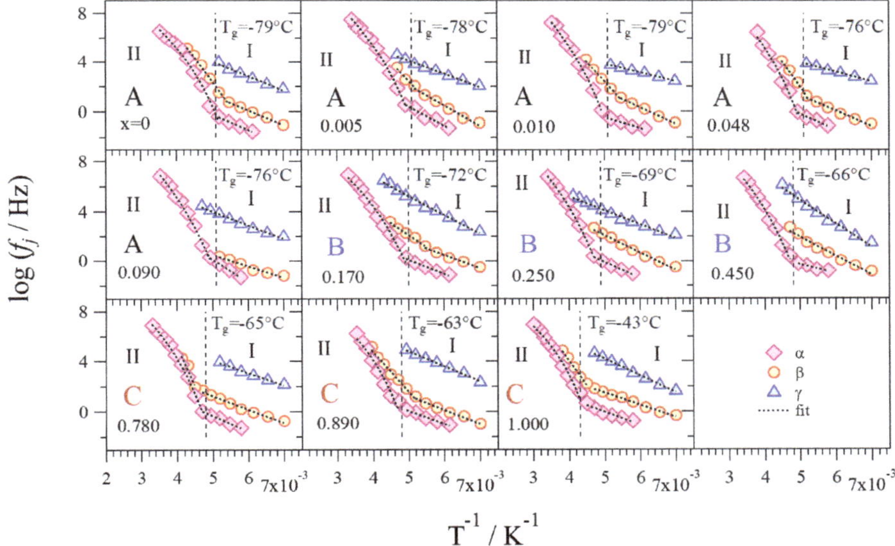

Figure 11. Log f_j vs T^{-1} curves of each dielectric relaxation event for every sample. Regions I and II are divided by the T_g of the sample.

For all the samples, the α dielectric relaxation is observed in both Region I, where it follows an Arrhenius-like behavior, and Region II, where a VTFH behavior is detected. Thus, only at high temperatures (Region II) is this relaxation assisting the long-range migration of Li^+ through the segmental motions of the glycerol aggregates. It is also observed that, at temperatures higher than the T_g, $\Delta\varepsilon_\alpha$ is the only dielectric strength that gets values higher than 10 (Figure S8), indicating a lower interaction between α dipole moments [19]. As elsewhere described [32], β relaxation shows an Arrhenius-like behavior in the whole temperature range, indicating that it is responsible for the migration of Li^+ through hopping events between different glycerol molecules. The dielectric strength of the β relaxation is lower than that of the α event. Finally, the γ dielectric relaxation mode is present only at $T < T_g$, and it shows an Arrhenius-like dependence on T^{-1}, similarly to β. The corresponding dielectric strength ($\Delta\varepsilon_\gamma$) reaches the lowest values among all the relaxations.

2.4.3. Activation Energies

The activation energy (E_a) associated to each conduction pathway is calculated by fitting the profiles of conductivity and dielectric relaxation frequency values vs T^{-1} shown in Figures 9 and 11, respectively. The obtained results are summarized in Figure 12 and Tables S1 and S2 of the Supplementary Materials.

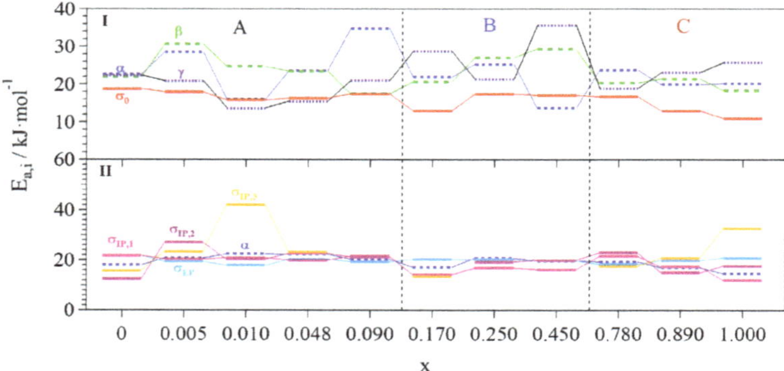

Figure 12. Activation energies calculated for each polarization phenomenon and relaxation event of every sample. The plot is divided into a low-temperature region (Region I, $T < T_g$) and a high-temperature region (Region II, $T > T_g$).

It is observed that at $T < T_g$ (Region I) and at low concentrations of Li^+ ($0.005 \leq x \leq 0.090$), E_a of σ_0 is very similar to the activation energy of the γ dielectric relaxation. Thus, for these samples and in this temperature range, the local fluctuations of the dipole moments of glycerol terminal hydroxyl functionalities are modulating the overall conductivity. As the content of Li^+ in the electrolytes is raised ($x \geq 0.170$), there is not a clear correlation between the activation energy of σ_0 and that of any relaxation. This allows us to conclude that in Region I the glycerol relaxations are not assisting the migration process of Li^+. In Region II, E_a of σ_k, and, in particular, E_a of σ_{EP}, perfectly overlap with the activation energy of α dielectric relaxation at all x values with a difference of no more than 4 kJ·mol^{-1}. Hence, independently from which coordination geometry glycerol molecules are assuming, the diffusion of conformational states between cluster aggregates is modulating the overall lithium ion conduction mechanism. Moreover, the activation energy of this relaxation is very low, always lower than 23 kJ·mol^{-1}. This indicates that the formation and dissolution mechanism of hydrogen bonds in clusters of glycerol molecules, required for the activation of α relaxation, is very efficient in long-range charge migration events.

2.4.4. Diffusion Coefficients and Average Charge Migration Distance

Additional insights into the conduction mechanism and the charge transfer processes occurring in the electrolytes are achieved by evaluating the temperature dependence of the diffusion coefficient (D), and the average charge migration distance ($\langle r \rangle$). D is calculated by means of the Nernst–Einstein equation (Equation (5)) [39]:

$$D_{\sigma_k,i} = \frac{RT_i\sigma_k}{Z_k^2 C_k F^2} \quad (5)$$

where R is the gas constant, T_i is the temperature at which D is calculated, σ_k is the conductivity of the *k*th polarization, Z_k and C_k are the charge and concentration of the *k*th species exchanged during the σ_k polarization, and F is the Faraday constant. The relationship between the diffusion coefficient D, associated with each polarization, and the dielectric relaxation events is derived from the Stokes–Einstein equation. The result is Equation (6):

$$D_{\sigma_k,i,j} = \frac{(k_b T_i)^{2/3}}{6\pi^{2/3}\tau_j^{1/3}\eta_i^{2/3}} \tag{6}$$

where k_b is the Boltzmann constant, τ_j is the relaxation time of the jth dielectric relaxation, and η is the viscosity of the system. The latter is approximated with the temperature-dependent viscosity of glycerol found in the literature [40]. The analysis of Equation (6) reveals that there is a linear correlation between the logarithm of the diffusion coefficient and the logarithm of the dielectric relaxation frequency. Results are shown in Figure S9 of the Supplementary Materials. The correlation between the diffusion coefficient and the relaxation frequencies at different temperatures and concentrations of Li^+ allows for the determination of the relationships between polarization phenomena and dielectric relaxation events. A relationship occurs when a linear dependence of the logarithm of the diffusion coefficient on the logarithm of the dielectric relaxation frequency is observed. Furthermore, this linear dependence has to be superimposable to the ideal one. As can be observed, all the polarizations are influenced by α dielectric relaxation, while β is correlated only to σ_{EP} and $\sigma_{IP,1}$. The γ dielectric relaxation does not influence any migration process. Focusing on the dependence on the lithium concentration it is observed that, at low Li^+ content (Group A), experimental and ideal data fit very well for both α and β dielectric relaxations. As the content of Li^+ is increased (Group B), experimental data overlap even better with the theoretical values, especially for the α relaxation, indicating that a high migration efficiency is present. Hence, it is expected that these electrolytes exhibit a high conductivity. Group C electrolytes present a good correlation between $D_{\sigma EP}$, $D_{\sigma IP1}$, $D_{\sigma IP2}$, and both α and β dielectric relaxations, even if $D_{\sigma IP1}$ and $D_{\sigma IP2}$ are a few $cm^2 \cdot s^{-1}$ lower than the ideal values.

The temperature dependence of the average charge migration distance (<r>) of the different percolation pathways is obtained from the Einstein–Smoluchowski equation (Equation (7)):

$$< r >_{k,i} = \sqrt{6D_{k,i}\tau_k} = \sqrt{\frac{6RT_i\sigma_k\tau_k}{Z_k^2 C_k F^2}} \tag{7}$$

where τ_k is the relaxation time of the σ_k polarization. Results are reported in Figure 13.

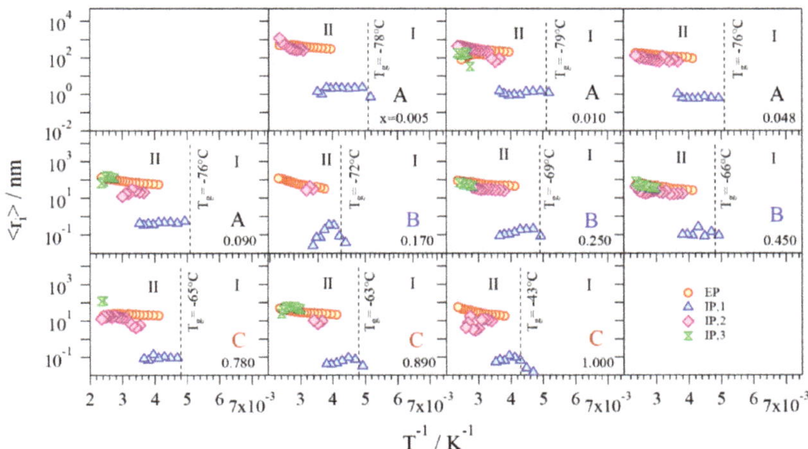

Figure 13. Average migration distance as a function of the temperature for σ_{EP}, $\sigma_{IP,1}$, $\sigma_{IP,2}$, and $\sigma_{IP,3}$.

It is observed that the migration distance increases as the temperature is raised. In Region I (below T_g), this migration is absent: Li^+ are entrapped between glycerol molecules, and a huge amount of energy is needed in order to activate the relaxation processes of the cluster aggregates and allow lithium ions to migrate. In Region II (above the T_g), $<r>$ is almost constant with the temperature for all the polarizations. On the contrary, the migration distance is affected by the different concentrations of Li^+ in the electrolytes: a decrease of one order of magnitude is observed going from the samples with x = 0.005 to x = 1.000. Indeed, when a high number of Li^+ ions able to be exchanged is present in the electrolyte, the dimension of the domains is smaller. The $<r_{EP}>$ values range from 10^2 to 10^1 nm; $<r_{EP}>$ is responsible for the long-range migration. On the contrary, $<r_{IP,1}>$ has the lowest values among all the polarizations, indicating that this phenomenon is responsible for the short-range migration. Moreover, this distance is equal to 1 nm, which is similar to the distance between two neighboring glycerol molecules. Thus, interdomain polarization 1 is attributed to the exchange of lithium ions between adjacent glycerol molecules. Similar to $<r_{EP}>$, $<r_{IP,2}>$ and $<r_{IP,3}>$ account for the long-range charge migration processes, and contribute with a higher intensity in electrolytes with a low concentration of Li^+. Furthermore, the average migration distances reflect the dimensions of the domains among which lithium ions are exchanged: IP,2 and IP,3 domains are more distant with respect to the domains involved in the IP,1 percolation pathways.

2.5. Conduction Mechanism

Broadband electrical spectroscopy measurements highlight (i) the presence of three different interdomain polarizations, associated with three Li^+ conduction pathways; and (ii) the existence of three types of nanodomains with different permittivities (ε_1, ε_2, and ε_3). Above the T_g, all the polarizations follow a VTF trend, indicating that they are all assisted by the dielectric relaxation modes of the glycerol aggregates, in particular the α relaxation. Each nanodomain is constituted by glycerol molecules in several coordination geometries—g1, g2.1, g2.2, and g3—which influence its permittivity value. Nanodomains 2 and 3 are predominantly present at high concentrations of Li^+ ($x \geq 0.170$), where $\sigma_{IP,2}$ and $\sigma_{IP,3}$ provide an important contribution to the long-range migration.

The $\sigma_{IP,1}$ polarization phenomenon is detected at $T_g \leq T \leq 10$ °C; it is responsible for a migration process IP,1 with a short range (ca. 1 nm). In this temperature range, the domains associated to $\sigma_{IP,1}$ are predominant and are constituted by large aggregates of crystalline glycerol molecules ("*Type 1 domains*") that (i) display the negatively charged $-O^-$ functionalities on their external surface and (ii) assume an ordered structure that allows the exchange of Li^+ between adjacent molecules. The formation of Type 1 domains allows for the explanation of the short-range migration pathways of Li^+ and of the occurrence of this polarization at low temperatures and high frequencies. The structure of Type 1 domains is rigid. Thus, the diffusion of conformational states along the glycerol aggregates is hindered. Hence, the conductivity of the electrolyte is low. An illustration elucidating the $\sigma_{IP,1}$ polarization phenomenon is shown in Figure 14a.

The $\sigma_{IP,2}$ polarization phenomenon is observed at temperatures higher than 10 °C and at lower frequencies with respect to $\sigma_{IP,1}$. $\sigma_{IP,2}$ is responsible for the long-range charge migration process of Li^+, reaching distances of ca. 1 μm. The domains associated with $\sigma_{IP,2}$ ("*Type 2 domains*") are characterized by the presence of a distribution of ordered aggregates with smaller dimensions in comparison with those associated with Type 1 domains. Indeed, $\sigma_{IP,2}$ occurs at temperatures higher than $\sigma_{IP,1}$. In $\sigma_{IP,2}$ polarization, Li^+ are exchanged between adjacent Type 2 domains; thus, a longer average charge migration distance is obtained. Moreover, the structure of Type 2 domains is more flexible with respect to that of Type 1 domains, improving the diffusion of conformational states within glycerol aggregates and facilitating the Li^+ migration process. The percolation of Li^+ following $\sigma_{IP,2}$ is depicted in Figure 14b.

The $\sigma_{IP,3}$ polarization appears at temperatures higher than 60 °C and at lower frequencies with respect to $\sigma_{IP,1}$ and $\sigma_{IP,2}$. IP,3, similarly to IP,2, accounts for a long-range migration process (ca. 1 μm). Nevertheless, it provides a lower contribution to the total conductivity if compared with IP,2. "*Type*

3 domains", which are associated with $\sigma_{IP,3}$, are composed of amorphous glycerol molecules with a distribution of coordination geometries falling between those observed in Type 1 and Type 2 domains. Indeed, this polarization is observed only at temperatures higher than 60 °C. The structure of Type 3 domains is very flexible. Thus, the migration of Li^+ is assisted by the diffusion of conformational states along glycerol molecules. The $\sigma_{IP,3}$ polarization mechanism is represented in Figure 14c.

Electrode polarization (σ_{EP}) starts to be the dominant polarization event at T > 10 °C in all the electrolytes. The presence of domains facilitates the long-range Li^+ transfer, which occurs via the exchange between different delocalization bodies (DBs) of the material [41]. A DB is a portion of the electrolyte with a defined volume that comprises Type 1, Type 2, and Type 3 domains. Since the exchange of Li^+ ions between the coordination sites present in each DB is very fast, Li^+ ions can be considered delocalized in each DB. Thus, in the proposed electrolytes, the long-range charge migration occurs as Li^+ ions are exchanged between different DBs as they come into contact, owing to conformational changes caused by the diffusion of conformational states of the glycerol molecules.

The analysis of the polarization phenomena and dielectric relaxation events allows for the proposal of the overall conduction mechanism shown in Figure 14d. In the overall conduction mechanism, the electrode polarization and all the three interdomain polarizations contribute to the diffusion of the Li^+ ions through different percolation pathways within the proposed electrolytes.

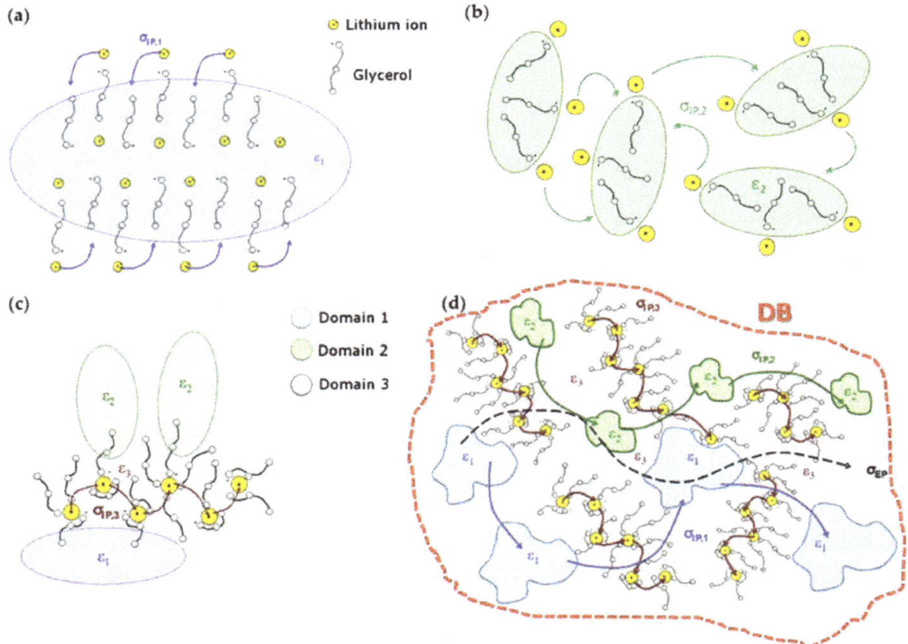

Figure 14. Proposed conduction mechanism for (**a**) $\sigma_{IP,1}$, (**b**) $\sigma_{IP,2}$, and (**c**) $\sigma_{IP,3}$; (**d**) overall proposed conduction mechanism. The delocalization body (DB) is highlighted with a red dotted line.

3. Materials and Methods

3.1. Materials

Glycerol (≥99%) was purchased from Carlo Erba, while lithium hydride (LiH, ≥99%) was obtained from Sigma-Aldrich (Milan, Italy). Glycerol was dried under vacuum at 60 °C for 8 days before use, and then transferred inside an Argon-filled glove-box (MBraun, less than 1 ppm of O_2 and H_2O). Then, 30 mL of dry glycerol were inserted into a 100 mL round flask, and 1.000 g of LiH were gradually added to the solution. The solution was vigorously stirred under vacuum at 50 °C in order to promote the reaction yielding lithium glycerolate, GlyLi (see Reaction I). GlyLi was diluted with different amounts of pristine glycerol, yielding 11 electrolytes with the general formula $GlyLi_x$, $0 \leq x \leq 1$.

3.2. Chemical and Structural Characterizations

The lithium content in all the electrolytes was determined via inductively coupled plasma atomic emission spectroscopy (ICP-AES) measurements. The ICP-AES analysis was carried out by digesting the samples in concentrated HNO_3. The emission line used to evaluate the Li concentration was 670.700 nm. The FT-IR spectra were collected using a Thermo Scientific (Monza, Italy) model Nicolet Nexus spectrometer, with a SPECAC Golden Gate single reflection diamond ATR cell. The spectrum was obtained as the average of 500 scans and with a resolution of 2 cm^{-1}. Micro-Raman analyses were carried out with a confocal DXR-Micro Raman (Thermo Scientific, Monza, Italy) spectrometer with an excitation laser operating at 532 nm. Samples were prepared inside an Ar-filled glove-box (see above), and were maintained under an inert atmosphere during the measurements. Both FT-IR and micro-Raman spectra were baseline subtracted and normalized with respect to the peaks centered at 1323 cm^{-1} (FT-IR) and 1253 cm^{-1} (micro-Raman), attributed to the superposition of the in-plane OH bending and CH wagging.

3.3. Thermal Characterizations

The thermal losses were evaluated by high-resolution thermogravimetric analyses (HR-TGA), using a high-resolution microbalance from TA Instruments (Sesto San Giovanni (MI), Italy) (model 2950). Measurements were carried out in the temperature range between 30 and 950 °C, under an N_2 atmosphere. The thermal transitions were studied via modulated differential scanning calorimetry (MDSC); measurements were carried out from −150 to 165 °C with a heating ramp of 3 $°C \cdot min^{-1}$.

3.4. Electric Response

Broadband electrical spectroscopy (BES) measurements were performed using an Alpha-A analyzer provided by Novocontrol Technologies (Montabaur, Germany). The complex conductivity ($\sigma^*(\omega)$) and permittivity spectra ($\varepsilon^*(\omega)$) were measured in the −130 to 150 °C temperature range at increments of 10 ± 0.5 °C. The measurements were carried out in the 10^{-1}–10^{7} Hz frequency range. Samples were loaded in a hermetically sealed homemade Teflon cell assembled in an Ar glove-box, and kept under a dry N_2 atmosphere during the measurements.

4. Conclusions

In this work, a new family of Li^+-conducting glass-forming polymer electrolytes is proposed. Vibrational spectroscopies (i.e., FT-IR and micro-Raman) allow for the interpretation of the coordination geometry assumed by glycerol molecules as a function of lithium concentration. High-resolution thermogravimetric analyses demonstrate that the proposed electrolytes are thermally stable up to 170 °C, while modulated differential scanning calorimetry measurements highlight that, as the content of Li^+ is increased, the T_g shifts to higher temperatures (from −78.51 °C for pristine glycerol to −42.95 °C for lithium glycerolate, GlyLi). Furthermore, the thermal behavior of the proposed electrolytes can be explained considering that glycerol molecules assume the coordination geometries hypothesized from vibrational results. Finally, broadband electrical spectroscopy results show that the highest conductivity values are achieved by $GlyLi_{0.25}$: 1.99×10^{-4} at 30 °C, higher than 10^{-6} $S \cdot cm^{-1}$ at −10 °C, and 1.55×10^{-2} $S \cdot cm^{-1}$ at 150 °C. Moreover, a detailed analysis of BES results allows us to propose a reasonable conduction mechanism for Li^+. In particular, it is shown that Li^+ ions are able to migrate between crystalline and amorphous domains thanks to the diffusion of conformational states of glycerol molecule aggregates. At temperatures below the T_g, the conduction mechanism occurs through "hopping" processes between adjacent glycerol molecules. On increasing the temperature, new polarization phenomena arise such as $\sigma_{IP,1}$, which is the principal event between T_g and 10 °C. At temperatures higher than 10 °C, the electrode polarization σ_{EP} becomes the dominant contribution to the overall conductivity. σ_{EP} consists of an exchange of lithium ions between different delocalization bodies (DBs), which results in a long-range migration of charge. In this temperature range, the contribution of $\sigma_{IP,2}$ polarization becomes relevant, and the flexibility obtained from glycerol molecules raises the conductivity of the proposed electrolytes. The best conductivity values obtained by the proposed electrolytes are promising in comparison with state-of-the-art polymer electrolytes for application in Li-ion batteries.

Supplementary Materials: The following are available online at http://www.mdpi.com/2313-0105/4/3/41/s1, Figure S1: FT-IR differential spectra. Figure S2: Intensity and wavenumber behavior of ν(OH), ν^{g3}(C–C–O), and ν^{g1}(C–C–O) FT-IR vibrations at different x (n_{Li}/n_{gly}) values. Figure S3: Lorentzian fitting of the micro-Raman spectrum of pure glycerol. Figure S4: Areal behavior of the micro-Raman peaks as a function of x. Figure S5: MDSC results of pristine glycerol, Gly (x = 0), and lithium glycerolate, GlyLi (x = 1.000), samples where the reversible and nonreversible heat flow contributions are separated. Figure S6: 3D tanδ surfaces of $GlyLi_x$ samples, with x = 0.005, 0.048, 0.090, 0.250, 0.450, 0.780, and 0.890. Figure S7: Representation of glycerol α, β, and γ dielectric relaxations. **Figure S8**: $\Delta\varepsilon_j$ vs T^{-1} curves of each dielectric relaxation event for every sample. Figure S9: Diffusion coefficient calculated for each polarization (σ_{EP}, $\sigma_{IP,1}$, and $\sigma_{IP,2}$ rows) as a function of the dielectric relaxation frequencies (f_γ, f_β, and f_α, columns). Table S1: Activation energies calculated for each polarization phenomenon. Table S2: Activation energies calculated for each relaxation event.

Author Contributions: V.D.N. conceived and designed the experiments; G.P. and S.T. performed the experiments; G.P., S.T., and K.V. analyzed the data; G.P., S.T., K.V., and V.D.N. discussed the results and agreed on the conclusions; G.P. and V.D.N. wrote the paper.

Funding: This work was supported by the "Budget Integrato per la Ricerca Interdipartimentale—BIRD 2016" of the University of Padova (protocol BIRD164837/16), and the "Centro Studi di Economia e Tecnica dell'Energia Giorgio Levi Cases" of the University of Padova. V.D.N. thanks the University Carlos III of Madrid for the "Cátedras de Excelencia UC3M-Santander" (Chair of Excellence UC3M-Santander).

Conflicts of Interest: The authors declare no conflict of interest.

References

1. Di Noto, V.; Lavina, S.; Giffin, G.A.; Negro, E.; Scrosati, B. Polymer electrolytes: Present, past and future. *Electrochim. Acta* **2011**, *57*, 4–13. [CrossRef]
2. Meyer, W.H. Polymer Electrolytes for Lithium-Ion Batteries. *Adv. Mater.* **1998**, *10*, 439–448. [CrossRef]
3. Deng, Y.; Eames, C.; Fleutot, B.; David, R.; Chotard, J.N.; Suard, E.; Masquelier, C.; Islam, M.S. Enhancing the Lithium Ion Conductivity in Lithium Superionic Conductor (LISICON) Solid Electrolytes through a Mixed Polyanion Effect. *ACS Appl. Mater. Interfaces* **2017**, *9*, 7050–7058. [CrossRef] [PubMed]
4. Tarascon, J.M.; Guyomard, D. New electrolyte compositions stable over the 0 to 5 V voltage range and compatible with the $Li_{1+x}Mn_2O_4$/carbon Li-ion cells. *Solid State Ion.* **1994**, *69*, 293–305. [CrossRef]
5. Canepa, P.; Dawson, J.A.; Sai Gautam, G.; Statham, J.M.; Parker, S.C.; Islam, M.S. Particle Morphology and Lithium Segregation to Surfaces of the $Li_7La_3Zr_2O_{12}$ Solid Electrolyte. *Chem. Mater.* **2018**, *30*, 3019–3027. [CrossRef]
6. Dawson, J.A.; Canepa, P.; Famprikis, T.; Masquelier, C.; Islam, M.S. Atomic-Scale Influence of Grain Boundaries on Li-Ion Conduction in Solid Electrolytes for All-Solid-State Batteries. *J. Am. Chem. Soc.* **2018**, *140*, 362–368. [CrossRef] [PubMed]
7. Bertasi, F.; Negro, E.; Vezzù, K.; Nawn, G.; Pagot, G.; Di Noto, V. Single-Ion-Conducting Nanocomposite Polymer Electrolytes for Lithium Batteries Based on Lithiated-Fluorinated-Iron Oxide and Poly(ethylene glycol) 400. *Electrochim. Acta* **2015**, *175*, 113–123. [CrossRef]
8. Bertasi, F.; Vezzù, K.; Giffin, G.A.; Nosach, T.; Sideris, P.; Greenbaum, S.; Vittadello, M.; Di Noto, V. Single-ion-conducting nanocomposite polymer electrolytes based on PEG400 and anionic nanoparticles: Part 2. Electrical characterization. *Int. J. Hydrogen Energy* **2014**, *39*, 2884–2895. [CrossRef]
9. Bertasi, F.; Vezzù, K.; Negro, E.; Greenbaum, S.; Di Noto, V. Single-ion-conducting nanocomposite polymer electrolytes based on PEG400 and anionic nanoparticles: Part 1. Synthesis, structure and properties. *Int. J. Hydrogen Energy* **2014**, *39*, 2872–2883. [CrossRef]

10. Bonino, F.; Croce, F.; Panero, S. Electrochemical characterization of an ambient temperature rechargeable Li battery based on low molecular weight polymer electrolyte. *Solid State Ion.* **1994**, *70–71*, 654–657. [CrossRef]
11. Pagot, G.; Bertasi, F.; Vezzù, K.; Nawn, G.; Pace, G.; Nale, A.; Di Noto, V. Correlation between Properties and Conductivity Mechanism in Poly(vinyl alcohol)-based Lithium Solid Electrolytes. *Solid State Ion.* **2018**, *320*, 177–185. [CrossRef]
12. Howell, F.S.; Moynihan, C.T.; Macedo, P.B. Electrical Relaxations in Mixtures of Lithium Chloride and Glycerol. *Bull. Chem. Soc. Jpn.* **1984**, *57*, 652–661. [CrossRef]
13. Adam, G.; Gibbs, J.H. On the Temperature Dependence of Cooperative Relaxation Properties in Glass-Forming Liquids. *J. Chem. Phys.* **1965**, *43*, 139–146. [CrossRef]
14. Lide, D.R.; Milne, G.W.A. *Handbook of Data on Organic Compounds*, 3rd ed.; CRC Press: Boca Raton, FL, USA, 1994; ISBN 9780849304453.
15. Van Duyne, R.P.; Reilley, C.N. Low-Temperature Electrochemistry II. Evaluation of Rate Constants and Activation Parameters for Homogeneous Chemical Reactions Coupled to Charge Transfer. *Anal. Chem.* **1972**, *44*, 153–158. [CrossRef]
16. Macedo, P.B.; Moynihan, C.T.; Bose, R. Role of Ionic Diffusion in Polarization in Vitreous Ionic Conductors. *Phys. Chem. Glasses* **1972**, *13*, 171–179.
17. Bastiansen, O. Intra-Molecular Hydrogen Bonds in Ethylene Glycol, Glycerol, and Ethylene Chlorohydrin. *Acta Chem. Scand.* **1949**, *3*, 415–421. [CrossRef]
18. Chelli, R.; Procacci, P.; Cardini, G.; Della Valle, R.G.; Califano, S. Glycerol condensed phases part I. A molecular dynamics study. *Phys. Chem. Chem. Phys.* **1999**, *1*, 871–877. [CrossRef]
19. Di Noto, V.; Vittadello, M.; Greenbaum, S.G.; Suarez, S.; Kano, K.; Furukawa, T. A New Class of Lithium Hybrid Gel Electrolyte Systems. *J. Phys. Chem. B* **2004**, *108*, 18832–18844. [CrossRef]
20. Chelli, R.; Procacci, P.; Cardini, G.; Califano, S. Glycerol condensed phases Part II.A molecular dynamics study of the conformational structure and hydrogen bonding. *Phys. Chem. Chem. Phys.* **1999**, *1*, 879–885. [CrossRef]
21. Chelli, R.; Gervasio, F.L.; Gellini, C.; Procacci, P.; Cardini, G.; Schettino, V. Density Functional Calculation of Structural and Vibrational Properties of Glycerol. *J. Phys. Chem. A* **2000**, *104*, 5351–5357. [CrossRef]
22. Zhao, Y.P.; Zhang, J.W.; Zhao, C.C.; Du, Z.Y. Tetrahedrally coordinated lithium(I) and zinc(II) carboxylate-phosphinates based on tetradentate 2-carboxyethyl(phenyl)phosphinate ligand. *Inorg. Chim. Acta* **2014**, *414*, 121–126. [CrossRef]
23. Liddel, U.; Becker, E.D. Infra-red spectroscopic studies of hydrogen bonding in methanol, ethanol, and t-butanol. *Spectrochim. Acta* **1957**, *10*, 70–84. [CrossRef]
24. Lin-Vien, D.; Colthup, N.B.; Fateley, W.G.; Grasselli, J.G. *The Handbook of Infrared and Raman Characteristic Frequencies of Organic Molecules*; Academic Press: Cambridge, MA, USA, 1991; ISBN 978-0-12-451160-6.
25. Mendelovici, E.; Frost, R.L.; Kloprogge, T. Cryogenic Raman spectroscopy of glycerol. *J. Raman Spectrosc.* **2000**, *31*, 1121–1126. [CrossRef]
26. Perova, T.S.; Christensen, D.H.; Rasmussen, U.; Vij, J.K.; Nielsen, O.F. Far-infrared spectra of highly viscous liquids: Glycerol and triacetin (glycerol triacetate). *Vib. Spectrosc.* **1998**, *18*, 149–156. [CrossRef]
27. Zietlow, J.P.; Cleveland, F.F.; Meister, A.G. Substituted methanes. III. Raman spectra, assignments, and force constants for some trichloromethanes. *J. Chem. Phys.* **1950**, *18*, 1076–1080. [CrossRef]
28. Dou, B.; Dupont, V.; Williams, P.T.; Chen, H.; Ding, Y. Thermogravimetric kinetics of crude glycerol. *Bioresour. Technol.* **2009**, *100*, 2613–2620. [CrossRef] [PubMed]
29. Hong, J.L.; Zhang, X.H.; Wei, R.J.; Wang, Q.; Fan, Z.Q.; Qi, G.R. Inhibitory effect of hydrogen bonding on thermal decomposition of the nanocrystalline cellulose/poly(propylene carbonate) nanocomposite. *J. Appl. Polym. Sci.* **2014**, *131*. [CrossRef]
30. Angell, C.A. Entropy and fragility in supercooling liquids. *J. Res. Natl. Inst. Stand. Technol.* **1997**, *102*, 171–181. [CrossRef] [PubMed]
31. Claudy, P.; Jabrane, S.; Létoffé, J.M. Annealing of a glycerol glass: Enthalpy, fictive temperature and glass transition temperature change with annealing parameters. *Thermochim. Acta* **1997**, *293*, 1–11. [CrossRef]
32. Di Noto, V.; Giffin, G.A.; Vezzù, K.; Piga, M.; Lavina, S. *Broadband Dielectric Spectroscopy: A Powerful Tool for the Determination of Charge Transfer Mechanisms in Ion Conductors, Solid State Proton Conductors: Properties and Applications in Fuel Cells*; John Wiley & Sons: Chichester, UK, 2012; pp. 109–183.

33. Kitajima, S.; Bertasi, F.; Vezzù, K.; Negro, E.; Tominaga, Y.; Di Noto, V. Dielectric relaxations and conduction mechanisms in polyether-clay composite polymer electrolytes under high carbon dioxide pressure. *Phys. Chem. Chem. Phys.* **2013**, *15*, 16626–16633. [CrossRef] [PubMed]
34. Tammann, G.; Hesse, W. Die Abhängigkeit der Viscosität von der Temperatur bie unterkühlten Flüssigkeiten. *Z. Anorg. Allg. Chem.* **1926**, *156*, 245–257. [CrossRef]
35. Dhatarwal, P.; Choudhary, S.; Sengwa, R.J. Electrochemical performance of Li^{+}-ion conducting solid polymer electrolytes based on PEO–PMMA blend matrix incorporated with various inorganic nanoparticles for the lithium ion batteries. *Compos. Commun.* **2018**, *10*, 11–17. [CrossRef]
36. Di Noto, V.; Piga, M.; Giffin, G.A.; Lavina, S.; Smotkin, E.S.; Sanchez, J.Y.; Iojoiu, C. Influence of anions on proton-conducting membranes based on neutralized nafion 117, triethylammonium methanesulfonate, and triethylammonium perfluorobutanesulfonate. 2. electrical properties. *J. Phys. Chem. C* **2012**, *116*, 1370–1379. [CrossRef]
37. Lunkenheimer, P.; Schneider, U.; Brand, R.; Loid, A. Glassy dynamics. *Contemp. Phys.* **2000**, *41*, 15–36. [CrossRef]
38. Kudlik, A.; Benkhof, S.; Blochowicz, T.; Tschirwitz, C.; Rössler, E. The dielectric response of simple organic glass formers. *J. Mol. Struct.* **1999**, *479*, 201–218. [CrossRef]
39. Di Noto, V.; Vittadello, M.; Yoshida, K.; Lavina, S.; Negro, E.; Furukawa, T. Broadband dielectric and conductivity spectroscopy of Li-ion conducting three-dimensional hybrid inorganic-organic networks as polymer electrolytes based on poly(ethylene glycol) 400, Zr and Al nodes. *Electrochim. Acta* **2011**, *57*, 192–200. [CrossRef]
40. Cheng, N.-S. Formula for the Viscosity of a Glycerol−Water Mixture. *Ind. Eng. Chem. Res.* **2008**, *47*, 3285–3288. [CrossRef]
41. Di Noto, V.; Piga, M.; Giffin, G.A.; Vezzù, K.; Zawodzinski, T.A. Interplay between mechanical, electrical, and thermal relaxations in nanocomposite proton conducting membranes based on nafion and a $[(ZrO_2)\cdot(Ta_2O_5)_{0.119}]$ core-shell nanofiller. *J. Am. Chem. Soc.* **2012**, *134*, 19099–19107. [CrossRef] [PubMed]

Article

Ion Transport in Solvent-Free, Crosslinked, Single-Ion Conducting Polymer Electrolytes for Post-Lithium Ion Batteries

Clay T. Elmore †, Morgan E. Seidler †, Hunter O. Ford, Laura C. Merrill, Sunil P. Upadhyay, William F. Schneiderand Jennifer L. Schaefer *

Department of Chemical and Biomolecular Engineering, University of Notre Dame, Notre Dame, IN 46556, USA; Clay.T.Elmore.5@nd.edu (C.T.E.); Morgan.E.Seidler.10@nd.edu (M.E.S.); hford1@nd.edu (H.O.F.); Laura.C.Merrill.16@nd.edu (L.C.M.); Sunil.Upadhyay@va.gov (S.P.U.); wschneider@nd.edu (W.F.S.)

* Correspondence: Jennifer.L.Schaefer.43@nd.edu; Tel.: +1-574-631-5114

† These authors contributed equally to this work.

Received: 14 May 2018; Accepted: 6 June 2018; Published: 7 June 2018

Abstract: Solvent-free, single-ion conducting electrolytes are sought after for use in electrochemical energy storage devices. Here, we investigate the ionic conductivity and how this property is influenced by segmental mobility and conducting ion number in crosslinked single-ion conducting polyether-based electrolytes with varying tethered anion and counter-cation types. Crosslinked electrolytes are prepared by the polymerization of poly(ethylene glycol) diacrylate (PEGDA), poly(ethylene glycol) methyl ether acrylate, and ionic monomers. The ionic conductivity of the electrolytes is measured and interpreted in the context of differential scanning calorimetry and Raman spectroscopy measurements. A lithiated crosslinked electrolyte prepared with $PEG_{31}DA$ and (4-styrenesulfonyl)(trifluoromethanesulfonyl)imide (STFSI) monomers is found to have a lithium ion conductivity of 3.2×10^{-6} and 1.8×10^{-5} S/cm at 55 and 100 °C, respectively. The percentage of unpaired anions for this electrolyte was estimated at about 23% via Raman spectroscopy. Despite the large variances in metal cation–STFSI binding energies as predicted via density functional theory (DFT) and large variations in ionic conductivity, STFSI-based crosslinked electrolytes with the same charge density and varying cations (Li, Na, K, Mg, and Ca) were estimated to all have unpaired anion populations in the range of 19 to 29%.

Keywords: polymer electrolyte; single-ion conducting; ionic conductivity; Raman spectroscopy

---

1. Introduction

Advanced energy storage devices are desired to support next-generation consumer electronics and defense technologies, facilitate widespread electric vehicle adoption, and increase the penetration of renewably-generated electricity into the grid. Worldwide, research in battery materials and technology seeks to address these needs. Post-lithium ion batteries, including those based on lithium metal and other more abundant materials, are under active investigation. Next-generation battery electrolytes that allow for improved safety, maintained or longer device lifetimes, and that support post-lithium ion platforms are highly desired [1–6].

The most common Li-ion battery electrolytes that are used today are based on organic aprotic solvents, such as organic carbonates, and used in conjunction with a microporous polymer separator or imbibed in a polymer gel. These electrolytes exhibit high Li-ion conductivity and support an adequate solid-electrolyte-interface (SEI) formation, but they present many practical disadvantages, such as being volatile and flammable [1,5,7,8]. One class of electrolytes that are seen as a possible alternative are polymer electrolytes (PE). PEs offer many advantages, including improved thermal and electrochemical stability [2,9,10]. Inorganic solid electrolytes typically have even better thermal stability, but they do not offer the same flexibility as PEs, which may inhibit their use in large-format devices [5,11].

A specific class of PEs of recent increased research interest is the single-ion conducting polymer electrolyte (SIPE) [12]. The typical molecular structure of the SIPE for a lithium-ion battery is a lithiated ionomer with poly(ethylene oxide) (PEO) functionality and tethered anions [13–21]. SIPEs can have very high oxidative stabilities and support higher charge/discharge rates than polymer electrolytes of similar conductivities and non-unity transference numbers [16,22]. The free anion in a traditional polymer electrolyte with a mobile salt is usually the limiting factor for electrochemical stability with the lowest oxidative stability in the electrolyte. In the SIPE, the anion is fixed to the polymer backbone and does not migrate to the anode where it would otherwise degrade.

The ionic conductivity of SIPEs is heavily influenced by the chemical nature of their anion. Anions with delocalized electron densities have been shown to enable high ionic conductivities, because the delocalized charge promotes dissociation of the ion pair and increases free cation concentration in the solvating polymer (such as the PEO chain) [13,18,23–26]. One such highly delocalized tethered anion is (4-styrenesulfonyl)(trifluoromethanesulfonyl)imide (STFSI). An even more delocalized tethered anion, 4-styrenesulfonyl)(trifluoromethyl(S-trifluoromethylsulfonylimino)sulfonyl)imide] (SsTFSI), has been shown to enable record ionic conductivity of a dry SIPE (1.35×10^{-4} S/cm), however, its synthesis is complex [13].

Furthermore, the macromolecular architecture also has a large impact on the performance of a SIPE. Longer amorphous chain segments allow fast segmental dynamics and, therefore, faster cation mobility; however, if the PEO segments are too long, the chains will crystallize resulting in drastically reduced cation mobility [15,18]. Electrolytes incorporating polymerized poly(ethylene glycol) methyl ether acrylate (PEGMA) have been shown to support high ionic conductivities over a wide temperature range. One such system is the random PEGMA-STFSI copolymer; unfortunately, these polymers are sticky gels and not convenient for battery fabrication [17]. Previous work has been done with poly(ethylene glycol) diacrylate (PEGDA) to create crosslinked, polymer electrolyte films. An important potential advantage to these crosslinked polymer electrolytes is their ability to be blade-coated directly on electrode sheets. However, PEGDA is available commercially at low cost only for short chain lengths (M_n~700 g/mol), and crosslinked electrolytes based on this short PEGDA and without liquid plasticizer exhibit quite low ionic conductivities. Potential solutions to this challenge include the use of longer chain PEGDA and the copolymerization of PEGDA and PEGMA [27–31].

The most researched SIPE systems to date have been lithium-ion based SIPEs. Next-generation batteries based on other metals, such as Na, Mg, and Al, however, are under increasing investigation due to the increased widespread abundance of these elements. Lithium reserves are located in sporadic locations around the globe [32–35]. In contrast, Mg, Na, and K can be harvested commercially from ocean water [32–34,36]. The commercial availability of Mg, combined with its high energy capacity, makes Mg a great candidate for an alternative platform to Li for high energy density batteries [36–38]. Other metals, including Na, have lower specific energy capacity, but are under active investigation for grid-level energy storage platforms [39–48].

Herein, we report on the ion transport properties of solvent-free, crosslinked SIPEs for post-lithium ion batteries. The use of longer chain PEGDA and the copolymerization of PEGDA and PEGMA are investigated in tandem with tethered anion chemistry and counter-cation type. The specific effects of segmental mobility and ion pairing on the conductivity are quantified via differential scanning calorimetry, Raman spectroscopy, and density functional theory (DFT) binding energy calculations.

2. Results and Discussion

Crosslinked electrolytes were prepared as described in the methods by the polymerization of PEGDA of varying molecular weights (M_n = 700, 1000, 1600, 2150, and 4700 g/mol), PEGMA (M_n = 480 g/mol and 750 g/mol), and an ionic monomer, followed by ion-exchange and drying.

The molecular structures of the electrolytes are shown in Figure 1. The ion content was held constant at a ratio of 1:30 moles charge:moles ethylene oxide (EO) for all cases described here. This ion content was chosen as we have found several poly(ethylene) glycol-based ionomers to have an optimum conductivity in an intermediate temperature range at this ion content. These include $PEG_{31}DA$-x-STFSILi and $PEG_{31}DA$-x-SSLi crosslinked ionomers as further studied here for temperatures of 40–90 °C, as well as the bottlebrush copolymer PEG_9MA-ran-STFSILi.

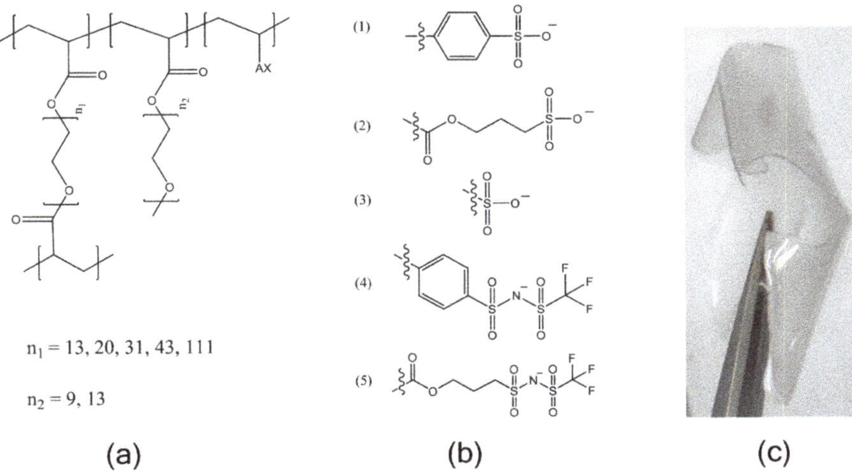

Figure 1. Chemical structures of (**a**) crosslinked ionomeric electrolytes, where A is the tethered anion and X is the counter-cation, and (**b**) the tethered anions, referred to as follows: (1) SS, (2) APS, (3) VS, (4) STFSI, and (5) APTFSI. A photograph of a typical crosslinked electrolyte is shown in (**c**).

Lithiated single-ion conducting electrolytes were prepared with five different common tethered anions: styrene sulfonate (SS), vinyl sulfonate (VS), acrylate-propylsulfonate (APS), (4-styrenesulfonyl)(trifluoromethanesulfonyl)imide (STFSI), and acrylate-propyl (trifluoromethanesulfonyl) imide (APTFSI). The TFSI-derivatives have been shown to be more dissociable, resulting in higher ionic conductivities than the sulfonates, but the chemical structure of the polymerizable group alters the rigidity of the ionomer which also affects ionic conductivity. This experiment was undertaken so that the relative effects of ion pair dissociation and segmental mobility on the ion transport could be compared over the temperature range of −20 to 100 °C. Figure 2a displays the ionic conductivity of the electrolytes based on the varying bound anions, while Figure 2b displays the measured glass transition temperatures (T_gs) of the electrolytes and the DFT predicted dissociation energies of the bound anion-Li ion pairs. Lower T_gs are correlated with higher chain segmental mobilities, and lower dissociation energies are correlated with higher mobile cation concentrations. Comparison of Figure 2a,b suggests that the degree of tethered anion dissociability most influenced the conductivity at high temperatures, whereas segmental mobility has an increased influence at low temperatures. Notice that at high temperatures, approaching 100 °C, the electrolytes containing TFSI-based anions exhibit ionic conductivities of over one order of magnitude greater than that of the sulfonate-based electrolytes; this difference is much diminished at low temperatures. This indicates that the ion pair dissociation and ion mobility have different activation energies [19]. This effect is also made apparent when comparing styrenic and non-styrenic monomers. The influence of higher glass transition temperatures and, therefore, reduced segmental mobility for the styrene-based ionic monomers results in a more appreciable decrease in conductivity (σ) at lower temperatures. Specifically, $T_{g,\mathrm{APTFSI}} < T_{g,\mathrm{STFSI}}$ and $T_{g,\mathrm{VS}} < T_{g,\mathrm{APS}} < T_{g,\mathrm{SS}}$; at −20 °C, $\sigma_{\mathrm{APTFSI}} > \sigma_{\mathrm{STFSI}}$ and $\sigma_{\mathrm{VS}} > \sigma_{\mathrm{APS}} > \sigma_{\mathrm{SS}}$. This same trend does not hold at higher temperatures.

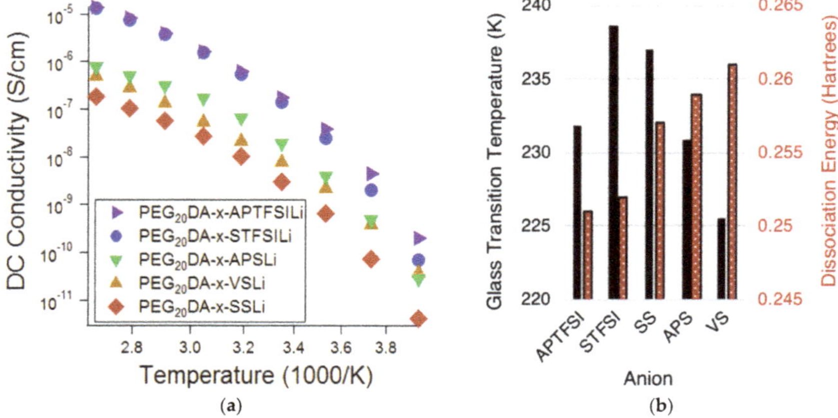

Figure 2. (**a**) DC conductivity from −20 to 100 °C for electrolytes containing the five different anions, and (**b**) corresponding experimental T_g values for these electrolytes and DFT calculated dissociation energies for ion pairs ALi, where A is the anion on the y-axis.

Of the investigated tethered anion types, electrolytes containing APTFSI exhibited the highest ionic conductivity over the measured temperature range with SFTSI electrolytes exhibiting just slightly lower conductivities. The remaining investigations reported here employed electrolytes based on STFSI, as despite the slightly reduced conductivities, we found that this monomer was easier to prepare with high purity.

We also found that the differences in binding energies between the sulfonate and modified anions resulted in different mesoscale structures, as observed with small angle and wide angle X-ray scattering (SAXS-WAXS, Figure 3). Markedly less structure was observed in the intermediate angle region, corresponding to nanometer length scales, for the crosslinked ionomers containing STFSI than for those containing SS. The correlation at point q_a, corresponding to a characteristic distance of about 5 nm, is clearly present in $PEG_{31}DA$-x-SSLi, but absent for $PEG_{31}DA$-x-STFSILi, which instead displays a minor shoulder at point q_b, a real-space distance of about 2.5 nm. Likewise, $PEG_{13}DA$-x-SSLi clearly displays a correlation length at point q_c, a characteristic distance of about 2 nm, which is again much less prominent for $PEG_{13}DA$-x-STFSILi. For these materials we hypothesize that evolution of a correlation length is indicative of the formation of ionic aggregates, a process which is partly governed by the degree of ion pair dissociation. With a low degree of dissociation, as is observed in the case of the SS anion, ionic charges are screened due to anion-cation pairing. This allows for the micro-phase separation of the polymers into PEGDA-rich and ionic unit-rich domains, a process which is subject solely to the physical constraints of the network crosslinking. However, as the dissociation degree increases, a greater number of unpaired anions exist in the crosslinked network. The tethered negative charges of the unpaired anions mitigate aggregation via electrostatic repulsion, resulting in a more uniform distribution of scatterers.

As the network crosslinker chain length increases, the ability of the polymer to accommodate greater aggregation increases, which results in the longer characteristic distances between ionic aggregates in the case of $PEG_{31}DA$-x-SSLi relative to $PEG_{13}DA$-x-SSLi. In addition, longer crosslinkers facilitate aggregation in the case of the $PEG_{31}DA$-x-STFSILi, as the enhanced conformational freedom of the $PEG_{31}DA$ allows for the exclusion of unpaired anions while facilitating some degree of micro-phase separation of paired ionic groups.

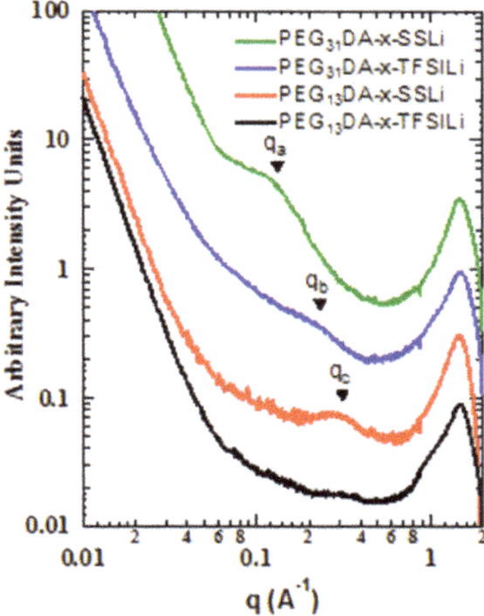

Figure 3. Small and wide angle X-ray scattering data on select crosslinked electrolytes at room temperature. Scattering vectors (q) corresponding with ionic aggregation are noted.

The influence of the network morphology of the lithiated crosslinked ionomers, namely the crosslinker length and presence branching, on the ionic conductivity was also investigated (Figure 4). Glass transition and melting transition temperatures, as well as ionic conductivity at 55 °C, as a function of crosslinked ionomer composition, is summarized in Table A1 in Appendix A. Ionic conductivity was found to be a strong function of PEGDA crosslinker length at ambient temperature, with $PEG_{31}DA$ resulting in the highest ionic conductivity at temperatures up to 40 °C. For temperatures in the range of 55 °C to 100 °C, the conductivity of the electrolytes prepared with crosslinkers in the range of $PEG_{20}DA$ to $PEG_{111}DA$ was similar. These results are attributed to the competing effects of segmental mobility, ion pairing, and crystallization. Glass transition temperatures decreased and, therefore, segmental mobility increased, with increasing crosslinker length. Crosslinkers of $PEG_{43}DA$ and longer were found to create networks that crystallized, resulting in a drop off in ionic conductivity at temperatures below the amorphous-crystalline transition. Crosslinkers of $PEG_{31}DA$ and $PEG_{20}DA$, while semi-crystalline in the monomer form, produced crosslinked networks that remain amorphous at all temperatures. Finally, we suspect that there is an effect of tethered anion proximity that results in increased ion pairing for the long crosslinkers. On average, there are more adjacent tethered anions with increased crosslinker length for materials with the same overall ion content. Anions adjacent along the chain are more likely to have polarizability volume overlap, which results in a lower degree of cation dissociation and therefore a lower effective mobile ion number [49,50]. These combined effects explain the non-monotonic relationship of ionic conductivity and network crosslinker length at various temperatures.

The addition of PEGMA combs was found to only modestly improve the ionic conductivity of the electrolytes, if at all. Representative data is shown in Figure 4b, which shows the effect of varying the ratio of $PEG_{9}MA$ to $PEG_{13}DA$ in the range of 0:1 to 4:1, where ionic conductivity is improved only in limited cases. We also investigated the addition of $PEG_{9}MA$ combs to networks based on $PEG_{20}DA$, as well as the addition of $PEG_{14}MA$ to networks based on $PEG_{31}DA$, and found similar results. Thus, we conclude that greater enhancements in conductivity due to chain branching are likely only when the networks approaches a loosely crosslinked bottlebrush, wherein there are nearly all PEGMA combs and a limited number of PEGDA crosslinkers. With the network fabrication methods utilized here and described in the experiment, we were unable to realize a soft solid of this morphology.

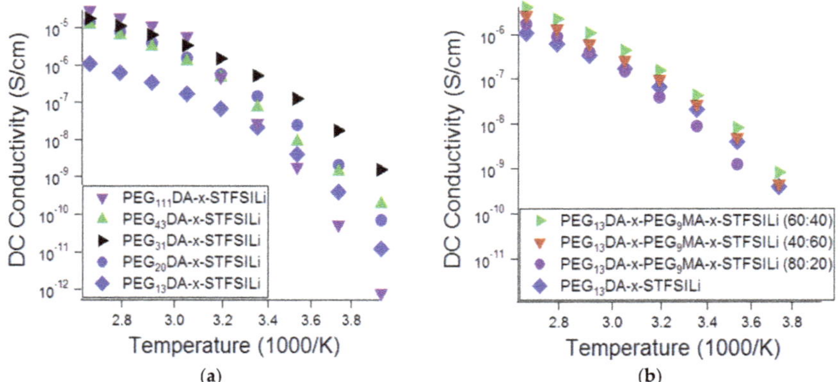

Figure 4. DC ionic conductivity from −20 to 100 °C for crosslinked electrolytes with (**a**) varying PEG crosslinker length and (**b**) varying ratios of crosslinker PEGDA and side-chain PEGMA.

The lithiated crosslinked electrolyte $PEG_{31}DA$-x-$PEG_{14}MA$-x-STFSILi was employed in a symmetric lithium metal coin cell and galvanostatically cycled to confirm its ability to reversibly strip and plate lithium metal. This electrolyte does effectively cycle lithium metal, as shown in Figure 5, with a small, but observable, increase in resistance over 150 h.

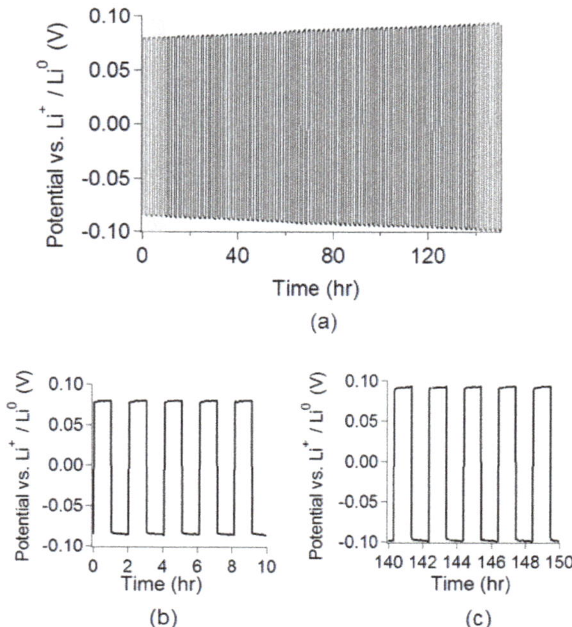

Figure 5. Galvanostatic cycling of a Li/$PEG_{31}DA$-x-$PEG_{14}MA$-x-STFSILi/Li coin cell at 50 μA/cm^2 at 70 °C: (**a**) time period of 0 to 150 h, with shaded areas displayed on a larger scale in (**b**,**c**).

With an eye toward post-lithium ion battery chemistries based on more abundant elements, crosslinked electrolytes were exchanged to other metal cations using aqueous ion-exchange solutions of other metal chloride salts. The metal cation–STFSI binding energies were predicted using DFT calculations and compared to the percentage of unpaired anions as measured using Raman spectroscopy and the ionic conductivity. As shown in Figure 6, for electrolytes based on Li, Na, K, Al, Mg, and Ca, the variation in the ionic conductivity due to cation type is about two orders of magnitude at 100 °C and over three orders of magnitude at −20 °C. Across these temperatures, lithiated electrolytes were the most conductive, while calcinated electrolytes were the least conductive. Surprising, aluminated electrolytes were found to have lower ionic conductivity than those with alkali (+1) cations, but higher ionic conductivity than the alkaline earth (+2) cations. Elemental analysis revealed that these electrolytes contain 1.5 times the aluminum predicted based upon $Al(STFSI)_3$ complexation, thus, there are chloride anions remaining. The resultant aluminum chloride complexes (could include $AlCl_2^+$, $AlCl_4^-$, etc.) have faster transport rates in the polyether ionomer than the bare divalent cations Mg^{2+} and Ca^{2+}. The low ionic conductivity of the hard divalent cations Mg^{2+} and Ca^{2+} in poly(ethylene oxide) has been documented in other works and is attributed to the divalent cation acting as a crosslinking point, strongly linking adjacent chains [51,52]. The conductivity of the Na and K containing electrolytes is close to that of the lithiated electrolyte at high temperatures, thus, we suggest that these electrolytes be further considered for use in elevated temperature Na and K battery systems.

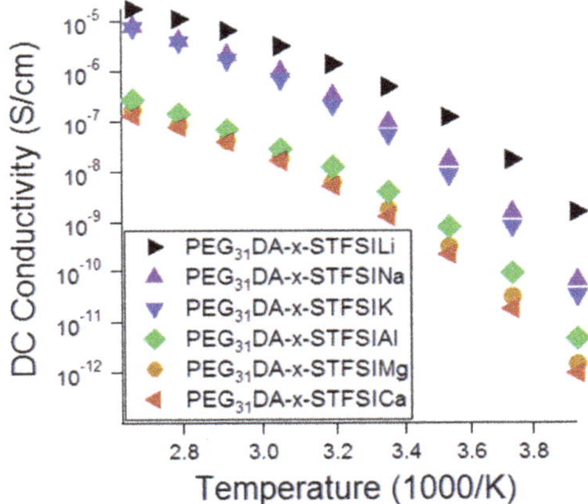

Figure 6. DC ionic conductivity from −20 to 100 °C for crosslinked electrolytes with varying cations.

Raman spectroscopy was used to further characterize the electrolytes to evaluate the relative influences of free cation number and cation mobility on ionic conductivity. First, theoretical Raman spectra were generated using DFT in order to support the experimental Raman peak identification. A directory summarizing the DFT inputs/outputs is available as Supplemental Materials. A DFT prediction using the PBE method with a 6311++G(d,p) basis for the LiSTFSI molecule showed that the vibrational mode that is associated with the S-N-S bonds expanding and contracting symmetrically was significantly affected by coordination with a cation. The shift from the free to the paired state was ~18 wavenumbers for coordination with Li^+; specifically, the free peak was predicted at 691 cm^{-1} while the peak was predicted at 673 cm^{-1} for the lithiated anion. Furthermore, DFT predictions showed no significant Raman modes within 80 cm^{-1} of this characteristic STFSI peak, indicating that this peak would be relatively undisturbed by the surrounding signal from the STFSILi molecule. Shown in Figure 7 is the experimental Raman spectra for $PEG_{31}DA$-x-STFSILi, and Figure 8 displays an enlarged view of the region of interest. The peak at ~732 cm^{-1} is attributed to the unpaired STFSI- population, whereas the peak at ~746 cm^{-1} is attributed to the STFSILi ion pair. Note that these modes occur at wavenumbers similar to that found for the common TFSI- anion, $CF_3SO_2NSO_2CF_3^-$ [26,53,54].

In no case could a peak be identified at a sufficient signal-to-noise ratio that corresponded to the cationic triplet, $STFSILi_2^+$. Notably, the Raman spectra of the electrolytes with varying metal cations appeared quite similar in this region in that two distinct peaks were apparent that we attribute to free STFSI and paired STFSI. For divalent metal cations, non-negligible populations of the neutral triplet are anticipated, i.e., $STFSI_2Mg$. The peak associated with this neutral triplet is predicted to occur close to that $STFSIMg^+$, but could not be unambiguously identified at the level of noise present. Thus, in the following estimations of free and paired STFSI populations for the Mg and Ca-exchanged electrolytes, both the charged and neutral triplets are accounted for as part of the paired STFSI population. Unfortunately, the Al-exchanged electrolyte PEG_{31}DA-x-STFSIAl was opaque and, thus, the collected signal intensity was not high enough to produce a meaningful Raman spectrum for this composition.

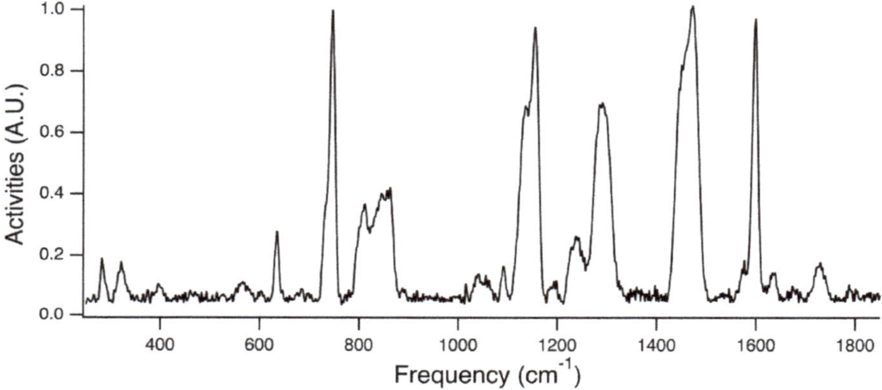

Figure 7. Experimental Raman spectrum of PEG_{31}DA-x-STFSILi.

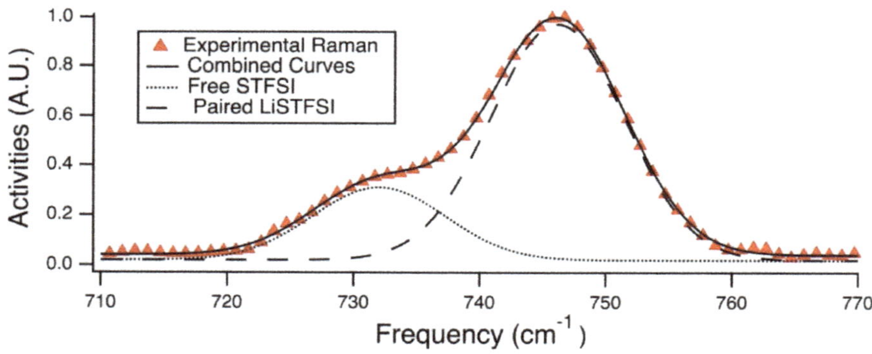

Figure 8. Peak fitting of PEG_{31}DA-x-STFSILi Raman spectrum to determine percentages of paired and unpaired (free) anions. The spectra for films of other compositions appeared quite similar.

Raman spectra were fit to provide estimations of the percentage of unpaired anions, see Figure 9a. For monovalent cations, the percentage of unpaired anions is exactly equivalent to the percentage of unpaired cations. Increasing ion size at equivalent valency, corresponding to decreased charge density, resulted in increased unpaired anion fractions. The highest unpaired fraction measured was 29 ± 2% for K-exchanged electrolytes, compared with the lowest unpaired fraction of about 19% for Mg-exchanged electrolytes, a relative decrease of less than 35%. This was found surprising given the three-fold difference in STFSI-cation binding energies as predicted from DFT, shown in Figure 9b. However, for divalent cations, this binding energy is that predicted for STFSI⁻ and X^{2+}; the binding energy for $STFSI^-$ and $STFSIX^+$ is expected to be considerably lower. Hence, for multivalent cations, the percentage of unpaired anions is not equivalent to the percentage of unpaired anions. Thus, with this analysis we are unable to unambiguously attribute the low divalent cation conductivity observed in Figure 6 to low numbers of mobile cations or low conducting cation mobility in the matrix.

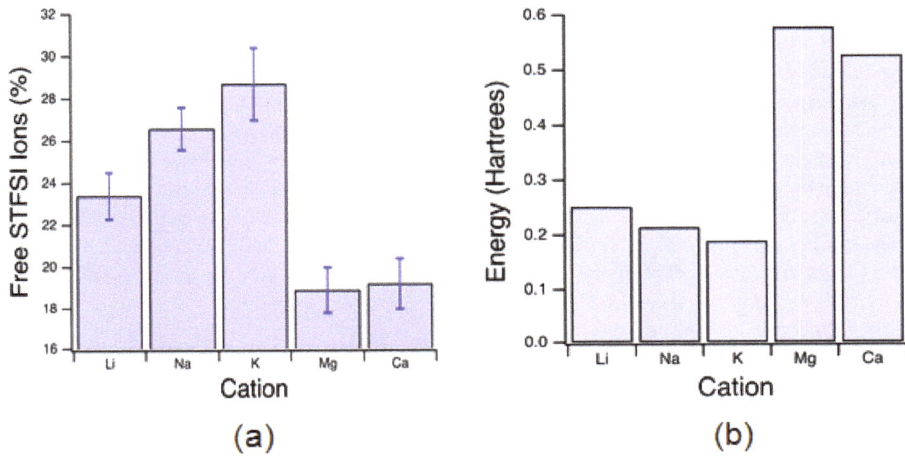

Figure 9. (**a**) Measured free ion percentages from Raman spectroscopy for PEG_{31}DA-x-STFSIX electrolytes where X is the variable cation, and (**b**) DFT predicted dissociation energies for STFSIX ion pairs where X is the variable cation.

The number of free alkali cations, as determined from the total cation mass fraction and material density, may be used in concert with the conductivity equation and the Nernst-Einstein equation to provide an estimate of the conducting cation mobility for the case where the conducting species is the free, monatomic cation. The ionic conductivity, σ, is the product of the conducting ion number (n), ion charge (q), and ion mobility (μ):

$$\sigma = nq\mu \tag{1}$$

The ion mobility is related to the ion diffusivity, D, as follows:

$$D = \frac{\mu kT}{q} \tag{2}$$

where k is the Boltzmann constant and T is the temperature. Using the unpaired cation fraction of -23% for Li^+ as determined from Raman spectroscopy and a measured density of 1.28 g/mL, we estimate the Li^+ diffusivity in PEG_{31}DA-x-STFSILi as 6×10^{-24} cm^2/s at 20 °C. Na^+ and K^+ diffusivities at 20 °C are estimated as 5×10^{-25} and 3×10^{-25} cm^2/s, respectively, for PEG_{31}DA-x-STFSINa and PEG_{31}DA-x-STFSIK.

The Li^+ diffusivity value estimated here should in the future be compared to the ^{7}Li self-diffusion coefficient as measured via pulse-field gradient nuclear magnetic resonance spectroscopy (PFG-NMR) [55]. While several other works have used Raman spectroscopy to estimate the degree of TFSI anion coordination in electrolytes such as ionic liquids, we believe that this is the first effort to apply this approach to single-ion conducting polymers with tethered STFSI anions. A prior work used FTIR to estimate the degree of tethered sulfonate coordination in a different single-ion conducting polymer [50]. Few simple methods exist for assessing the populations of free and coordinated cations, and diffusion coefficient measurements via PFG-NMR require more expensive instrumentation and are limited to certain nuclei such as ^{7}Li and ^{1}H.

3. Conclusions

Several factors are found to influence the ionic conductivity of these PEGDA crosslinked single-ion conducting electrolytes, including crosslinker chain length and side-chains, anionic monomer proximity along the chain, tethered anion and counter-cation chemistry. We note that electrolytes prepared with PEGDA crosslinkers of at least 1600 g/mol exhibit higher ionic conductivities over a wide temperature range. The addition of PEGMA side-chains was found to improve conductivity in limited cases. The tethered anion chemistry was found to strongly influence ionic conductivity, but more so at elevated temperatures. Lithiated electrolytes were found to exhibit the highest ionic conductivities, but the Na-exchanged and K-exchanged electrolytes exhibited conductivity as nearly as high at 100 °C. The multivalent electrolytes were found to have significantly lower conductivities at all temperatures. It was demonstrated that percentages of unpaired STFSI anions may be quantified through the use of Raman spectroscopy measurements. This provides a measurement of the unpaired alkali cation population, but the multivalent cation states were not able to be quantified. We, therefore, suggest that future efforts in multivalent SIPEs be targeted toward characterization of cation states and solvation that ultimately dictate ionic conductivity.

4. Materials and Methods

4.1. Materials

The majority of chemicals were obtained from Sigma-Aldrich (St. Louis, MO, USA), including: poly(ethylene glycol) (PEG) of average molecular weights 1500, 2050, and 4600 g/mol, poly(ethylene glycol) methyl ether acrylate of average molecular weights 480 and 750 g/mol, poly(ethylene glycol) diacrylate of average molecular weight 700 g/mol, triethylamine, acryloyl chloride, thionyl chloride, oxalyl chloride, dichloromethane (DCM), tetrahydrofuran (THF), diethyl ether, anhydrous acetonitrile, dimethylformamide (DMF), methanol, pentane, 4-(dimethylamino)pyridine, lithium chloride, magnesium chloride, sodium chloride, aluminum chloride hexahydrate, calcium chloride, potassium carbonate, sodium bicarbonate, lithium hydride, hydrochloric acid, and 2-hydroxy-4′-(2-hydroxyethoxy)-2-methylpropiophenone (photoinitiator). Poly(ethylene glycol) diacrylate of average molecular weight 1000 g/mol was obtained from Polysciences (Warrington, PA, USA). Trifluoromethanesulfonamide was obtained from TCI America (Portland, OR, USA). All reagents were used as received.

4.2. Synthesis of Potassium 4-Styrenesulfonyl(trifluoromethylsulfonyl)imide (KSTFSI)

KSTFSI was synthesized via a previously reported procedure [18]. Oxalyl chloride (20 g, 157 mmol) and 0.55 mL DMF were added to 250 mL anhydrous acetonitrile and stirred for five hours under a nitrogen atmosphere to form the yellow Vilsmeier-Haack complex. To this solution, 25 g (121 mmol) of 4-styrenesulfonate sodium salt were added and the mixture was stirred for one day. The resultant NaCl salt was removed from the 4-styrene sulfonyl chloride solution via filtration. Separately, 51 mL of triethylamine, 18.04 g trifluoromethylsulfonamide (121 mmol), and 9% by weight 4-(dimethylamino)pyridine were added to 188 mL of anhydrous acetonitrile and allowed to dissolve under nitrogen. This solution was added dropwise to the 4-styrene sulfonyl chloride solution at 0 °C, and then stirred for 16 h. The final solution was concentrated via rotary evaporation, and the resultant solids were dissolved in 50 mL of dichloromethane. The dichloromethane solution was washed with 3×90 mL of 4% $NaHCO_3$ followed by 2×125 mL of 1 M hydrochloric acid. The washed organic solution was concentrated, with the acid monomer residue neutralized by the addition of one molar excess of K_2CO_3 dissolved in deionized (DI) water. The KSTFSI monomer was obtained by subsequent recrystallizations from DI water until confirmed pure via ^{1}H and ^{13}C NMR.

4.3. Synthesis of Lithium 1-[3-(Acryloyloxy)propylsulfonyl]-1-(trifluoromethane-sulfonyl)imide (LiAPTFSI)

LiAPTFSI was synthesized via a previously reported procedure [56]. Briefly, the purchased potassium 3-(acryloyloxy)propane-1-sulfonate was dried and then converted to 3-(chlorosulfonyl)propyl acrylate via reaction with excess thionyl chloride. Then, 3-(chlorosulfonyl)propyl acrylate was converted to triethyl ammonium 1-[3-(acryloyloxy)propylsulfonyl]-1-(trifluoromethane-sulfonyl)imide by reaction with trifluoromethanesulfonamide in the presence of trimethylamine. The monomer was then lithiated by reaction with lithium hydride (2.5 times excess) in THF. The product was recrystallized from dichloromethane, washed with pentane, and dried under high vacuum. The structure was confirmed via ^{1}H and ^{13}C NMR though the yield was low (21% for the final lithiation and purification steps).

4.4. Synthesis of Poly(ethylene glycol) Diacrylate (PEGDA)

PEGDA was synthesized in accordance with prior literature [57]. PEGs (M_n = 1500, 2050, and 4600 g/mol) were dissolved in DCM and reacted with triethylamine and 2.2 equivalents of acryloyl chloride in a dark environment at 0 °C (gradually increasing to room temperature) under a nitrogen atmosphere. The solution was concentrated via rotary evaporation to roughly 25% of the original volume, to which THF was added to precipitate triethylamine hydrochloride salts. The solution was then further concentrated to an oily residue, dissolved in 5 mL dichloromethane, and precipitated in chilled diethyl ether. The PEGDA monomer was collected via filtration and dried under vacuum for 24 h. Attachment of acrylate groups and purity was confirmed by ^{1}H NMR.

4.5. Preparation of Crosslinked Electrolytes

PEGDA of the required molecular weight was dissolved in DMF. The ionic monomer was added in the ratio of 30 PEGDA ethylene oxide units to one ionic monomer, and photoinitiator was added at the molar ratio of 0.01 mol/mol monomer. The mixture was exposed to UV radiation (UVC 515 Ultraviolet Multilinker, Ultra-Lum, Inc., Carson, CA, USA) and heated to 80 °C for the polymerization reaction. The polymerized films were then washed and ion-exchanged to the form of interest. First, the films were transferred to methanol, then to a 0.4 M aqueous ion exchange solution for 96 h with solution changes every 24 h and, finally, a pure water bath. The films were dried in the open atmosphere for 24 h, and finally dried in a glovebox under vacuum at 80 °C for 16 h.

4.6. Ionic Conductivity

Film conductivity was measured using a Broadband Dielectric/Impedance Spectrometer model Alpha A (Novocontrol Technologies, Montabaur, Germany) with symmetric brass electrodes. Measurements were taken at 15 °C intervals from −20 °C to 100 °C. AC voltage was set to 0.1 V and the frequency was swept from 10^7 Hz to 0.1 Hz.

4.7. Thermal Properties

The glass transition temperature (T_g) and melting temperature (T_m) were measured using differential scanning calorimetry (Mettler Toledo, DSC 1, Columbus, OH, USA) at a heating rate of 10 °C min^{-1} under a nitrogen atmosphere, on the second heating leg of a heat-cool-heat cycle.

4.8. DFT Calculations

Geometric optimizations and single point energy calculations of anion-cation systems were used to calculate dissociation energies were performed in accordance with the widely-used three-parameter Becke model with the Lee-Yang-Par modification (B3LYP) [58,59]. The basis set that was used for all of the optimizations and single point energy calculations was 6-311++G(d,p), a large basis set containing diffusive functions that help deal with the distributed charge in the system, especially the anions in question. Dissociation energy correlations between different tethered anions in the system were determined starting with an optimization of the geometry of a single anion and a single monatomic cation. The final geometry for the molecule was then used to obtain a structure for the anion of the system. A single point energy calculation with the same method and basis was then performed on the anion. Additionally, a single point energy of the monatomic cation was also calculated. The three energies that had been calculated were then used to calculate a dissociation energy as shown in Figures 2 and 9. Dissociation energies were also tested with a varying size of bases, and were shown to converge to the reported values with increasing basis size. Furthermore, cations were placed in many different positions for the start of optimization in order to ensure the finding of the global minimum in calculations. It should also be noted that in the salt systems studied in this paper originally contained a double bond that was used to polymerize the salt to the PEGDA system. This double bond was altered to a fully-hydrogenated single bond in the DFT calculations in order to simulate a structure more similar to the ones present in the studied polymers. Various starting geometries were tested in order to ensure the global minimum of geometric optimization was found.

The method illustrated in Figure 10 was extended to many different anion-cation systems to obtain relative dissociation energy predictions. For the dissociation energy calculations of lithium 3-sulfopropyl acrylate (LiAPS) and lithium 4-propylacrylatesulfonyl(trifluoromethylsulfonyl)imide (LiAPTFSI), the oxygen atoms in the acrylate group were substituted with fully hydrogenated carbon atoms. The effect of the acryl oxygen atoms to the dissociation energy of LiSPA and LiSPTFSI in the real PEGDA-containing system is substantially diluted at the charge:EO ratio of 1:30, as used here.

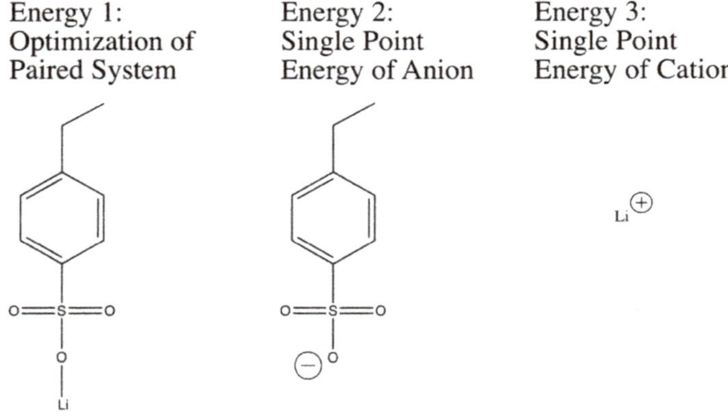

Figure 10. Schematic to describe system states used in DFT calculations to determine a relative dissociation energy for a lithium styrene sulfonate (LiSS) ion pair.

Theoretical Raman spectroscopy was also calculated using the 6-311++G(d,p) basis set and the generalized gradient approximation (GGA) method proposed by Perdew, Burke, and Ernzhof [60]. This GGA method will be referred to as "PBE" throughout the rest of the paper. This method was used because it is known to give similar results to B3LYP with a much smaller computational cost. The specific peak shifts that were studied in this paper corresponded to the difference in anion's vibration when it was paired with a monatomic cation, as opposed to when it was in the unpaired, "free", state. These shifts were determined by optimizing the tethered anion and calculating a Raman spectrum then optimizing the free anion and calculating Raman spectrum for comparison. Shifts of interest, which contained the same vibrational modes, were determined by using the Avogadro software to visualize the calculated vibrations [61]. All calculations were performed using the Gaussian09 computational software [62].

4.9. Raman Spectroscopy

Polymer films that were previously prepared in an argon box were then placed in a Teflon-sealed quartz cuvette inside the same glove box. The samples were flush to the quartz glass when they were taken out of the glove box. Raman Measurements were taken using a NRS-5000 Mirco-Raman Spectrometer (Jasco, Inc., Easton, MD, USA) with the 20× objective magnification. A 532 nm excitation laser was used as well as a resolution of 13.8 cm^{-1}. It should be noted that peak shapes were not significantly changed at a large range of resolutions (0.7 to 13.8 cm^{-1}), so the lowest resolution was used in order to increase the signal-to-noise ratio. The spectra were recorded every 1 cm^{-1}.

Population analysis was done via Gaussian curve fitting of these Raman spectra in the range of 710 cm^{-1} to 770 cm^{-1}. The peak at ~746 cm^{-1} corresponds to the paired STFSI anion, and the shoulder occurring at a lower wavenumber corresponds to the free STFSI anion. Using this knowledge, the Beer-Lambert law was applied to the area integrations for the two Gaussian curves that were fit to the these modes, shown in Figure 8, in order to determine the fraction of free STFSI anions. It should be noted that this calculation of free ions did not take into account any cation-anion configurations other than a single anion bound to one monatomic cation. Other geometric possibilities that could be present and have similar Raman activities. These geometries for monovalent cations include the triplet ($STFSILi_2$+), the negatively-charged coordination of two anions and one cation ($[STFSI]_2Li$-). For the divalent cations ($[STFSI]_2Mg$) is also a possibility. With relatively low signal-to-noise ratios, our analysis is somewhat restricted in determining which of these states are actually present in the electrolytes.

The error analysis that was performed on the free ion percentage calculations, shown in Figure 9, was done via Monte Carlo simulation. This was accomplished by taking a section of spectrum that was known to not have any signal and extracting an average random error to the known zero point. This random error was then introduced into the fitting Gaussian curves at random along the fitted curve. This new noisy curve was then fit using the same Gaussian curve fitting algorithm as the original data. The resulting two Gaussian curves were used to obtain a new population estimate. This population was stored, and the process was repeated 1000 times. A standard deviation was taken of all of the resulting population percentages, and a two standard deviation limit was placed on Figure 9 to depict a 95% confidence interval.

4.10. Galvanostatic Cycling

Li/PEG_{31}DA-x-PEG_{14}MA-x-STFSILi/Li coin cells, size 2032, were assembled in an argon glovebox. After a 24 h rest at room temperature and one hour rest at 70 °C, the coin cell was galavanostatically cycled at a rate of 50 $\mu A/cm^2$ for 2 h for each complete cycle at 70 °C using a BTS3000 battery tester (Neware, Shenzhen, China) with wiring into a gravity oven (VWR, Radnor, PA, USA).

4.11. Small Angle and Wide Angle X-Ray Scattering (SAXS-WAXS)

SAXS-WAXS measurements were obtained using the Advanced Photon Source (beamline 12-ID-B, operated by the Chemical and Materials Science Group) at Argonne National Lab, with an X-ray beam wavelength of 0.9322 Å (photon energy of 13.3 keV). The q-range was calibrated using a silver behenate standard, and the sample to detector distance was maintained at 2013 mm for the SAXS detector and 430.15 mm for the WAXS detector. Samples were prepared by loading the polymer films into glass or quartz capillary tubes (Charles Supper Company, Natick, MA, USA) and which were then sealed with wax under an argon atmosphere. Two-dimensional (2D) spectra were obtained with an exposure time of one second, collected with a Pilatus 2M camera (Dectris, Baden-Dättwil, Switzerland) for SAXS and Pilatus 300 for WAXS, and converted to 1D spectra via azimuthal integration. Spectra were corrected for transmission and background, and analyzed using the Igor Pro-Irena SAS package to display an arbitrary intensity vs scattering vector q, for:

$$q = 4\pi/\lambda * \sin(\theta) \tag{3}$$

where λ is the X-ray wavelength and θ is half of the scattering angle, 2θ.

4.12. Elemental Analysis

Inductively coupled plasma optically emitting spectroscopy (ICP-OES) was completed using a PerkinElmer Optima 8000 to quantify the amount of metal in the polymer films. Samples were digested by refluxing for 16 h in concentrated nitric acid (70%), then diluted to 5% nitric acid using 18 MΩ water. Calibration standards were made from 1000 ppm standards in 2% nitric acid (TraceCERT from Sigma Aldrich).

Supplementary Materials: The following are available online at http://www.mdpi.com/2313-0105/4/2/28/s1, directory of DFT inputs/outputs.

Author Contributions: Conceptualization: J.L.S.; methodology: C.T.E., M.E.S., H.O.F., and S.P.U.; software: C.T.E. and W.F.S.; validation: M.E.S., H.O.F., and J.L.S.; formal analysis: C.T.E. and M.E.S.; investigation: C.T.E., M.E.S., H.O.F., and L.C.M.; resources: W.F.S. and J.L.S.; data curation: C.T.E., H.O.F.; writing—original draft preparation: C.T.E., M.E.S., H.O.F., L.C.M., and J.L.S.; writing—review and editing: W.F.S. and J.L.S.; visualization: C.T.E., M.E.S., H.O.F., and J.L.S.; supervision: S.P.U.,W.F.S., and J.L.S.; project administration: J.L.S.; and funding acquisition: J.L.S.

Funding: This research was funded by the National Science Foundation (USA) through grant number CBET-1706370.

Acknowledgments: M.S. gratefully acknowledges financial support from ND Energy. The authors gratefully acknowledge use of the University of Notre Dame Materials Characterization Facility, Magnetic Resonance Facility, Center for Environmental Science and Technology, and Center for Research Computing. This research used resources of the Advanced Photon Source, a U.S. Department of Energy (DOE) Office of Science User Facility operated for the DOE Office of Science by Argonne National Laboratory under contract no. DE-AC02-06CH11357.

Conflicts of Interest: The authors declare no conflict of interest. The funding sponsors had no role in the design of the study; in the collection, analyses, or interpretation of data; in the writing of the manuscript; or in the decision to publish the results.

Appendix A

Table A1. Monomer ratios, thermal transition data, and DC ionic conductivity for crosslinked electrolytes of select compositions.

Composition	Ratio [1]	T_g (°C) [2]	T_m (°C) [4]	σ, [5] 55 °C (S/cm)
$PEG_{13}DA$-x-SSLi	2.3/0/1	−28.4	*n.d.*	3.3×10^{-9}
$PEG_{13}DA$-x-PEG_9MA-x-SSLi (80:20)	1.8/0.7/1	−37.5	*n.d.*	5.5×10^{-9}
$PEG_{13}DA$-x-PEG_9MA-x-SSLi (60:40)	1.4/1.3/1	−46.9	*n.d.*	9.5×10^{-10}
$PEG_{13}DA$-x-STFSILi	2.3/0/1	−17.8	*n.d.*	1.7×10^{-7}
$PEG_{13}DA$-x-PEG_9MA-x-STFSILi (80:20)	1.8/0.7/1	−23.4	*n.d.*	1.5×10^{-7}
$PEG_{13}DA$-x-PEG_9MA-x-STFSILi (60:40)	1.4/1.3/1	−36.3	*n.d.*	4.3×10^{-7}
$PEG_{20}DA$-x-SSLi	1.5/0/1	−36.2	*n.d.*	2.6×10^{-8}
$PEG_{20}DA$-x-VSLi	1.5/0/1	−47.6	*n.d.*	5.5×10^{-8}
$PEG_{20}DA$-x-APSLi	1.5/0/1	−42.3	*n.d.*	1.7×10^{-7}
$PEG_{20}DA$-x-APTFSILi	1.5/0/1	−41.3	*n.d.*	1.6×10^{-6}
$PEG_{20}DA$-x-STFSILi	1.5/0/1	−34.4	*n.d.*	1.6×10^{-6}
$PEG_{31}DA$-x-STFSILi	1/0/1	−38.5	*n.d.*	3.2×10^{-6}
$PEG_{31}DA$-x-STFSINa	1/0/1	−36.9	*n.d.*	1.0×10^{-6}
$PEG_{31}DA$-x-STFSIK	1/0/1	−43.5	30.8	6.8×10^{-7}
$PEG_{31}DA$-x-STFSIMg	1/0/1	−35.4	*n.d.*	1.8×10^{-8}
$PEG_{31}DA$-x-STFSICa	1/0/1	−42.0	*n.d.*	1.6×10^{-8}
$PEG_{31}DA$-x-STFSIAl	1/0/1	−42.5	*n.d.*	2.8×10^{-8}
$PEG_{43}DA$-x-STFSILi	0.7/0/1	−45.6	28.4	1.2×10^{-6}
$PEG_{111}DA$-x-STFSILi	0.3/1/1	*n.d.* [3]	41.8	5.7×10^{-6}

[1] PEGDA/PEGMA/Anion monomer number ratio; [2] glass transition temperature, accurate to ±1 °C; [3] *n.d.* = not detectable; [4] melting temperature, accurate to ±1 °C; [5] DC ionic conductivity at 55 °C.

References

1. Blomgren, G.E. The Development and Future of Lithium Ion Batteries. *J. Electrochem. Soc.* **2017**, *164*, A5019–A5025. [CrossRef]
2. Ngai, K.S.; Ramesh, S.; Ramesh, K.; Juan, J.C. A review of polymer electrolytes: Fundamental, approaches and applications. *Ionics (Kiel)* **2016**, *22*, 1259–1279. [CrossRef]
3. Wang, Y.; Chen, R.; Chen, T.; Lv, H.; Zhu, G.; Ma, L.; Wang, C.; Jin, Z.; Liu, J. Emerging non-lithium ion batteries. *Energy Storage Mater.* **2016**, *4*, 103–129. [CrossRef]
4. Choi, J.W.; Aurbach, D. Promise and reality of post-lithium-ion batteries with high energy densities. *Nat. Rev. Mater.* **2016**, *1*, 16013. [CrossRef]
5. Kalhoff, J.; Eshetu, G.G.; Bresser, D.; Passerini, S. Safer Electrolytes for Lithium-Ion Batteries: State of the Art and Perspectives. *ChemSusChem* **2015**, *8*, 2154–2175. [CrossRef] [PubMed]
6. Leadbetter, J.; Swan, L.G. Selection of battery technology to support grid-integrated renewable electricity. *J. Power Sources* **2012**, *216*, 376–386. [CrossRef]
7. Manthiram, A.; Yu, X.; Wang, S. Lithium battery chemistries enabled by solid-state electrolytes. *Nat. Rev. Mater.* **2017**, *2*, 16103. [CrossRef]
8. Hassoun, J.; Scrosati, B. Review—Advances in Anode and Electrolyte Materials for the Progress of Lithium-Ion and beyond Lithium-Ion Batteries. *J. Electrochem. Soc.* **2015**, *162*, A2582–A2588. [CrossRef]
9. Xue, Z.; He, D.; Xie, X. Poly(ethylene oxide)-based electrolytes for lithium-ion batteries. *J. Mater. Chem. A* **2015**, *3*, 19218–19253. [CrossRef]
10. Shaplov, A.S.; Marcilla, R.; Mecerreyes, D. Recent Advances in Innovative Polymer Electrolytes based on Poly(ionic liquid)s. *Electrochim. Acta* **2015**, *175*, 18–34. [CrossRef]
11. Fergus, J.W. Ceramic and polymeric solid electrolytes for lithium-ion batteries. *J. Power Sources* **2010**, *195*, 4554–4569. [CrossRef]
12. Zhang, H.; Li, C.; Piszcz, M.; Coya, E.; Rojo, T.; Rodriguez-Martinez, L.M.; Armand, M.; Zhou, Z. Single lithium-ion conducting solid polymer electrolytes: Advances and perspectives. *Chem. Soc. Rev.* **2017**, *46*, 797–815. [CrossRef] [PubMed]
13. Ma, Q.; Zhang, H.; Zhou, C.; Zheng, L.; Cheng, P.; Nie, J.; Feng, W.; Hu, Y.S.; Li, H.; Huang, X.; et al. Single lithium-ion conducting polymer electrolytes based on a super-delocalized polyanion. *Angew. Chem.-Int. Ed.* **2016**, *55*, 2521–2525. [CrossRef] [PubMed]
14. Mazor, H.; Golodnitsky, D.; Peled, E.; Wieczorek, W.; Scrosati, B. A search for a single-ion-conducting polymer electrolyte: Combined effect of anion trap and inorganic filler. *J. Power Sources* **2008**. [CrossRef]
15. Rojas, A.A.; Inceoglu, S.; Mackay, N.G.; Thelen, J.L.; Devaux, D.; Stone, G.M.; Balsara, N.P. Effect of Lithium-Ion Concentration on Morphology and Ion Transport in Single-Ion-Conducting Block Copolymer Electrolytes. *Macromolecules* **2015**, *48*, 6589–6595. [CrossRef]
16. Bouchet, R.; Maria, S.; Meziane, R.; Aboulaich, A.; Lienafa, L.; Bonnet, J.-P.; Phan, T.N.T.; Bertin, D.; Gigmes, D.; Devaux, D.; et al. Single-ion BAB triblock copolymers as highly efficient electrolytes for lithium-metal batteries. *Nat. Mater.* **2013**, *12*, 452–457. [CrossRef] [PubMed]
17. Feng, S.; Shi, D.; Liu, F.; Zheng, L.; Nie, J.; Feng, W.; Huang, X.; Armand, M.; Zhou, Z. Single lithium-ion conducting polymer electrolytes based on poly[(4-styrenesulfonyl)(trifluoromethanesulfonyl)imide] anions. *Electrochim. Acta* **2013**, *93*, 254–263. [CrossRef]
18. Meziane, R.; Bonnet, J.P.; Courty, M.; Djellab, K.; Armand, M. Single-ion polymer electrolytes based on a delocalized polyanion for lithium batteries. *Electrochim. Acta* **2011**, *57*, 14–19. [CrossRef]
19. Fragiadakis, D.; Dou, S.; Colby, R.H.; Runt, J. Molecular mobility and Li+ conduction in polyester copolymer ionomers based on poly(ethylene oxide). *J. Chem. Phys.* **2009**, *130*, 064907. [CrossRef] [PubMed]
20. Dou, S.; Zhang, S.; Klein, R.J.; Runt, J.; Colby, R.H. Synthesis and characterization of poly(ethylene glycol)-based single-ion conductors. *Chem. Mater.* **2006**, *18*, 4288–4295. [CrossRef]
21. Watanabe, M.; Suzuki, Y.; Nishimoto, A. Single ion conduction in polyether electrolytes alloyed with lithium salt of a perfluorinated polyimide. *Electrochim. Acta* **2000**, *45*, 1187–1192. [CrossRef]
22. Doyle, M.; Fuller, T.F.; Newman, J. The importance of the lithium ion transference number in lithium/polymer cells. *Electrochim. Acta* **1994**, *39*, 2073–2081. [CrossRef]

23. Van Humbeck, J.F.; Aubrey, M.L.; Alsbaiee, A.; Ameloot, R.; Coates, G.W.; Dichtel, W.R.; Long, J.R. Tetraarylborate polymer networks as single-ion conducting solid electrolytes. *Chem. Sci.* **2015**, *6*, 5499–5505. [CrossRef] [PubMed]
24. Brandell, D.; Kasemägi, H.; Tamm, T.; Aabloo, A. Molecular dynamics modeling the Li-PolystyreneTFSI/PEO blend. *Solid State Ion.* **2014**, *262*, 769–773. [CrossRef]
25. Liang, S.; Choi, U.H.; Liu, W.; Runt, J.; Colby, R.H. Synthesis and Lithium Ion Conduction of Polysiloxane Single-Ion Conductors Containing Novel Weak-Binding Borates. *Chem. Mater.* **2012**, *24*, 2316–2323. [CrossRef]
26. Rey, I.; Johansson, P.; Lindgren, J.; Lassègues, J.C.; Grondin, J.; Servant, L. Spectroscopic and Theoretical Study of $(CF_3SO_2)_2N^-$ ($TFSI^-$) and $(CF_3SO_2)_2NH$ (HTFSI). *J. Phys. Chem. A* **1998**, *102*, 3249–3258. [CrossRef]
27. Paranjape, N.; Mandadapu, P.C.; Wu, G.; Lin, H. Highly-branched cross-linked poly(ethylene oxide) with enhanced ionic conductivity. *Polymer (Guildf)* **2017**, *111*, 1–8. [CrossRef]
28. Daigle, J.-C.; Vijh, A.; Hovington, P.; Gagnon, C.; Hamel-Pâquet, J.; Verreault, S.; Turcotte, N.; Clément, D.; Guerfi, A.; Zaghib, K. Lithium battery with solid polymer electrolyte based on comb-like copolymers. *J. Power Sources* **2015**, *279*, 372–383. [CrossRef]
29. Liang, Y.-H.; Wang, C.-C.; Chen, C.-Y. Synthesis and characterization of a new network polymer electrolyte containing polyether in the main chains and side chains. *Eur. Polym. J.* **2008**, *44*, 2376–2384. [CrossRef]
30. Sun, X.G.; Kerr, J.B. Synthesis and characterization of network single ion conductors based on comb-branched polyepoxide ethers and lithium bis(allylmalonato)borate. *Macromolecules* **2006**, *39*, 362–372. [CrossRef]
31. Cowie, J.M.G.; Spence, G.H. Novel single ion, comb-branched polymer electrolytes. *Solid State Ion.* **1999**, *123*, 233–242. [CrossRef]
32. Aurbach, D.; Gofer, Y.; Lu, Z.; Schechter, A.; Chusid, O.; Gizbar, H.; Cohen, Y.; Ashkenazi, V.; Moshkovich, M.; Turgeman, R.; et al. A short review on the comparison between Li battery systems and rechargeable magnesium battery technology. *J. Power Sources* **2001**, *97–98*, 28–32. [CrossRef]
33. Meshram, P.; Pandey, B.D.; Mankhand, T.R. Extraction of lithium from primary and secondary sources by pre-treatment, leaching and separation: A comprehensive review. *Hydrometallurgy* **2014**, *150*, 192–208. [CrossRef]
34. Swain, B. Recovery and recycling of lithium: A review. *Sep. Purif. Technol.* **2017**, *172*, 388–403. [CrossRef]
35. Grosjean, C.; Miranda, P.H.; Perrin, M.; Poggi, P. Assessment of world lithium resources and consequences of their geographic distribution on the expected development of the electric vehicle industry. *Renew. Sustain. Energy Rev.* **2012**, *16*, 1735–1744. [CrossRef]
36. Yoo, H.D.; Shterenberg, I.; Gofer, Y.; Gershinsky, G.; Pour, N.; Aurbach, D. Mg rechargeable batteries: An on-going challenge. *Energy Environ. Sci.* **2013**, *68*, 1754–5692. [CrossRef]
37. Song, J.; Sahadeo, E.; Noked, M.; Lee, S.B. Mapping the Challenges of Magnesium Battery. *J. Phys. Chem. Lett.* **2016**, *7*, 1736–1749. [CrossRef] [PubMed]
38. Muldoon, J.; Bucur, C.B.; Gregory, T. Quest for nonaqueous multivalent secondary batteries: Magnesium and beyond. *Chem. Rev.* **2014**, *114*, 11683–11720. [CrossRef] [PubMed]
39. Dunn, B.; Kamath, H.; Tarascon, J.-M. Electrical energy storage for the grid: A battery of choices. *Science* **2011**, *334*, 928–35. [CrossRef] [PubMed]
40. Slater, M.D.; Kim, D.; Lee, E.; Johnson, C.S. Sodium-Ion Batteries. *Adv. Funct. Mater.* **2013**, *23*, 947–958. [CrossRef]
41. Ellis, B.L.; Nazar, L.F. Sodium and sodium-ion energy storage batteries. *Curr. Opin. Solid State Mater. Sci.* **2012**, *16*, 168–177. [CrossRef]
42. Pan, H.; Hu, Y.-S.; Chen, L. Room-temperature stationary sodium-ion batteries for large-scale electric energy storage. *Energy Environ. Sci.* **2013**, *6*, 2338. [CrossRef]
43. Wang, W.; Luo, Q.; Li, B.; Wei, X.; Li, L.; Yang, Z. Recent progress in redox flow battery research and development. *Adv. Funct. Mater.* **2013**, *23*, 970–986. [CrossRef]
44. Divya, K.C.; Østergaard, J. Battery energy storage technology for power systems—An overview. *Electr. Power Syst. Res.* **2009**, *79*, 511–520. [CrossRef]
45. Larcher, D.; Tarascon, J.-M. Towards greener and more sustainable batteries for electrical energy storage. *Nat. Chem.* **2015**, *7*, 19–29. [CrossRef] [PubMed]
46. Su, D.; McDonagh, A.; Qiao, S.-Z.; Wang, G. High-Capacity Aqueous Potassium-Ion Batteries for Large-Scale Energy Storage. *Adv. Mater.* **2017**, *29*, 1604007. [CrossRef] [PubMed]

47. Chayambuka, K.; Mulder, G.; Danilov, D.L.; Notten, P.H.L. Sodium-Ion Battery Materials and Electrochemical Properties Reviewed. *Adv. Energy Mater.* **2018**, 1800079. [CrossRef]
48. Elia, G.A.; Marquardt, K.; Hoeppner, K.; Fantini, S.; Lin, R.; Knipping, E.; Peters, W.; Drillet, J.-F.; Passerini, S.; Hahn, R. An Overview and Future Perspectives of Aluminum Batteries. *Adv. Mater.* **2016**, *28*, 7564–7579. [CrossRef] [PubMed]
49. Choi, U.H.; Mittal, A.; Price, T.L.; Lee, M.; Gibson, H.W.; Runt, J.; Colby, R.H. Molecular Volume Effects on the Dynamics of Polymerized Ionic Liquids and their Monomers. *Electrochim. Acta* **2015**, *175*, 55–61. [CrossRef]
50. Wang, J.H.H.; Yang, C.H.C.; Masser, H.; Shiau, H.S.; O'Reilly, M.V.; Winey, K.I.; Runt, J.; Painter, P.C.; Colby, R.H. Ion States and Transport in Styrenesulfonate Methacrylic PEO9 Random Copolymer Ionomers. *Macromolecules* **2015**, *48*, 7273–7285. [CrossRef]
51. Plancha, M.J.C. Characterisation and modeling of multivalent polymer electrolytes. In *Polymer Electrolytes: Fundamentals and Applications*; Sequeira, C., Santos, D., Eds.; Woodhead Publishing Limited: Philadelphia, PA, USA, 2010; pp. 340–377.
52. Ikeda, S.; Mori, Y.; Furuhashi, Y.; Masuda, H. Multivalent cation conductive solid polymer electrolytes using photo-cross-linked polymers: II. Magnesium and zinc trifluoromethanesulfonate systems. *Solid State Ion.* **1999**, *121*, 329–333. [CrossRef]
53. Seo, D.M.; Boyle, P.D.; Sommer, R.D.; Daubert, J.S.; Borodin, O.; Henderson, W.A. Solvate structures and spectroscopic characterization of litfsi electrolytes. *J. Phys. Chem. B* **2014**, *118*, 13601–13608. [CrossRef] [PubMed]
54. Fujii, K.; Fujimori, T.; Takamuku, T.; Kanzaki, R.; Umebayashi, Y.; Ishiguro, S.I. Conformational equilibrium of bis(trifluoromethanesulfonyl) imide anion of a room-temperature ionic liquid: Raman spectroscopic study and DFT calculations. *J. Phys. Chem. B* **2006**, *110*, 8179–8183. [CrossRef] [PubMed]
55. Lafemina, N.H.; Chen, Q.; Colby, R.H.; Mueller, K.T. The diffusion and conduction of lithium in poly(ethylene oxide)-based sulfonate ionomers. *J. Chem. Phys.* **2016**, *145*, 114903. [CrossRef]
56. Shaplov, A.S.; Vlasov, P.S.; Armand, M.; Lozinskaya, E.I.; Ponkratov, D.O.; Malyshkina, I.A.; Vidal, F.; Okatova, O.V.; Pavlov, G.M.; Wandrey, C.; et al. Design and synthesis of new anionic polymeric ionic liquids with high charge delocalization. *Polym. Chem.* **2011**, *2*, 2609–2618. [CrossRef]
57. Käpylä, E.; Delgado, S.M.; Kasko, A.M. Shape changing photodegradable hydrogels for dynamic 3D cell culture. *ACS Appl. Mater. Interfaces* **2016**, *8*, 17885–17893. [CrossRef] [PubMed]
58. Becke, A.D. Density-functional exchange-energy approximation with correct asymptotic behavior. *Phys. Rev. A* **1988**, *38*, 3098–3100. [CrossRef]
59. Lee, C.; Yang, W.; Parr, R.G. Development of the Colle-Salvetti correlation-energy formula into a functional of the electron density. *Phys. Rev. B* **1988**, *37*, 785–789. [CrossRef]
60. Perdew, J.P.; Burke, K.; Ernzerhof, M. Generalized Gradient Approximation Made Simple. *Phys. Rev. Lett.* **1996**, *77*, 3865–3868. [CrossRef] [PubMed]
61. Hanwell, M.D.; Curtis, D.E.; Lonie, D.C.; Vandermeersch, T.; Zurek, E. Hutchison, G.R. Avogadro: An advanced semantic chemical editor, visualization, and analysis platform. *J. Cheminform.* **2012**, *4*, 17. [CrossRef] [PubMed]
62. *Gaussian 09 A.02*; Gaussian, Inc.: Wallingford, CT, USA, 2009; pp. 2–3.

Article

Formation and Stability of Interface between Garnet-Type Ta-doped $Li_7La_3Zr_2O_{12}$ Solid Electrolyte and Lithium Metal Electrode

Ryoji Inada *, Satoshi Yasuda, Hiromasa Hosokawa, Masaya Saito, Tomohiro Tojoand Yoji Sakurai

Department of Electrical and Electronic Engineering, Toyohashi University of Technology, 1-1 Tempaku-cho, Toyohashi, Aichi 4418580, Japan; yasuda@cec.ee.tut.ac.jp (S.Y.); hosokawa@cec.ee.tut.ac.jp (H.H.); saito@cec.ee.tut.ac.jp (M.S.); tojo@ee.tut.ac.jp (T.T.); sakurai@ee.tut.ac.jp (Y.S.)

* Correspondence: inada@ee.tut.ac.jp; Tel.: +81-532-44-6723

Received: 10 May 2018; Accepted: 6 June 2018; Published: 7 June 2018

Abstract: Garnet-type $Li_{7-x}La_3Zr_{2-x}Ta_xO_{12}$ (LLZT) is considered a good candidate for the solid electrolyte in all-solid-state lithium batteries because of its reasonably high conductivity around 10^{-3} S cm^{-1} at room temperature and stability against lithium (Li) metal with the lowest redox potential. In this study, we synthesized LLZT with a tantalum (Ta) content of 0.45 via a conventional solid-state reaction process and constructed a Li/LLZT/Li symmetric cell by attaching Li metal foils on the polished top and bottom surfaces of an LLZT pellet. We investigated the influence of heating temperatures and times on the interfacial charge-transfer resistance between LLZT and the Li metal electrode. In addition, the effect of the interface resistance on the stability for Li deposition and dissolution was examined using a galvanostatic cycling test. The lowest interfacial resistance of 25 Ω cm^2 at room temperature was obtained by heating at 175 °C (5 °C lower than the melting point of Li) for three to five hours. We confirmed that the current density at which the short circuit occurs in the Li/LLZT/Li cell via the propagation of Li dendrite into LLZT increases with decreasing interfacial charge transfer resistance.

Keywords: garnet; solid electrolyte; lithium metal; interface; charge-transfer resistance

1. Introduction

A high performance electrical energy storage device is a key technology in the sustainable development of a ubiquitous and clean energy society. Rechargeable lithium-ion batteries (LIBs), using graphite as the anode and organic liquid electrolyte and lithium transition-metal oxide as the cathode, were commercialized in 1991 and have since been widely used worldwide as a power source for mobile electronic devices, such as cell phones and laptop computers because of their high-energy density and reasonably good cycling performance. The development of middle- or large-scale LIBs has accelerated for use in automotive propulsion and stationary load-leveling for intermittent power generation from solar or wind energy [1–3]. However, a larger battery size creates more serious safety issues in LIBs; one of the main reasons for this being the increased amount of flammable organic liquid electrolytes.

All-solid-state lithium-ion batteries with a non-flammable inorganic Li^+ ion conductor as a solid electrolyte (SE) are expected to be the next generation of energy storage devices because of their high energy density, safety, and excellent cycle stability [4–6]. The SE materials should possess not only high Li^+ ion conductivity above one mS cm^{-1} at a room temperature but also chemical stability against electrode materials, air, and moisture. Oxide-based SE materials have rather lower conductivity and poor plasticity compared with sulfide-based materials, but they have other advantages such as their chemical stability and ease of handling [7,8].

Garnet-type Li^+ ion conducting oxide, $Li_7La_3Zr_2O_{12}$ (LLZ), has been widely investigated because of its good ionic conduction, excellent thermal performance, and wide electrochemical potential window [9]. LLZ has two different crystal phases: the cubic phase [9,10] and the tetragonal phase [11,12]. However, high conductivity above 10^{-4} S cm^{-1} at a room temperature range is only obtained in the former, densified by high temperature sintering. Partial substitution of the Zr^{4+} site in LLZ by other higher valence cations, such as Nb^{5+} [13,14], Ta^{5+} [15–23], W^{6+} [24,25], and Mo^{6+} [26], is effective at stabilizing the cubic phase, and the conductivity at room temperature is greatly improved to 1 mS cm^{-1} by controlling the dopants content and optimizing Li^+ concentration in the garnet framework. Although a solid-state battery with an Nb-doped LLZ as SE has already been demonstrated [13,27,28], a Ta-doped LLZ showed much better chemical stability against a Li metal electrode than when Nb-doped [29,30]. The other dopants, such as W^{6+}, Mo^{6+}, or Nb^{5+} in LLZ, could potentially become a redox center at relatively high potential against Li^+/Li [31].

The use of Li metal with extremely large gravimetric specific capacity (3860 mAh g^{-1}) with the lowest redox potential as an anode leads to the high energy density of a battery, but the formation of a solid-solid interface among garnet-type SE and Li metal electrodes is another challenging issue in achieving better electrochemical performance in solid-state batteries [32–34]. Many approaches have been introduced to reduce the interfacial charge-transfer resistance between garnet-type SE and Li, including the introduction of thin film layers of Au [35], Si [36], Ge [37], Al_2O_3 [38], and ZnO [39], or eliminating the secondary phases, such as LiOH and Li_2CO_3, by polishing the surface of SE and using multiple thermal treatments at specific temperatures before and after contact with Li [40–42]. However, a more simplified method to form the interface between garnet-type SE and the Li electrode would be preferable and further study is required of the relationship between the interfacial charge-transfer resistance and stability for Li deposition and dissolution reaction at the interface.

In this work, we synthesized a cubic garnet-type LLZT using a solid-state reaction process. An Li/LLZT/Li symmetric cell, created by attaching Li metal foil on the top and bottom surfaces of an LLZT pellet, was used to examine the influence of heating temperatures and times on the interfacial charge-transfer resistance between LLZT and Li. Moreover, the effect of the interface resistance on the stability of Li deposition and dissolution reactions at the Li/LLZT interface was systematically investigated using a galvanostatic cycling test.

2. Materials and Methods

2.1. Synthesis and Characterization of LLZT

LLZT with a Ta content (x) of 0.45 was synthesized using a conventional solid-state reaction process. Notably, we did not use an Al_2O_3 crucible for sample synthesis because Al^{3+} contamination from the crucible into the LLZT lattice during high temperature sintering may influence the Li^+ contents and, generally, this contamination level cannot be precisely controlled. The following were obtained from the Kojundo Chemical Laboratory (Saitama, Japan): stoichiometric amounts of $LiOH \cdot H_2O$ (99% with 10% excess was added to account for the evaporation of lithium at high temperatures), $La(OH)_3$ (99.99%), ZrO_2 (98%) and Ta_2O_5 (99.9%), which were then ground and mixed by planetary ball-milling (Nagao System, Planet M2-3F, Kawasaki, Japan) with zirconia balls and ethanol for 3 h in a zirconia pot, and then calcined at 900 °C for 6 h in air using a Pt-5% Au alloy crucible. The calcined powders were ground again by planetary ball-milling for 1 h, and then pressed into pellets at 300 MPa by cold isostatic pressing. Finally, the pellets were sintered at 1150 °C for 15 h in air using a Pt-5% Au alloy crucible. To minimize Li loss and the formation of secondary phases during the sintering process, the pellets were covered with the same mother powder.

The crystal phase of LLZT was evaluated by X-ray diffraction (XRD; Rigaku Multiflex, Tokyo, Japan) using Cu$K\alpha$ radiation (λ = 0.15418 nm), with a measurement range 2θ of 5–90° and a step interval of 0.004°. Scanning electron microscope (SEM) observation of the fractured surface microstructure of the sintered LLZT was performed using a scanning electron microscope (SU8000 Type II, HITACHI,

Tokyo, Japan).). The electrical conductivity of each LLZT was evaluated with alternating current (A/C) impedance measurements using a chemical impedance meter (3532-80, HIOKI, Ueda, Japan) at temperatures from 27 to 100 °C, frequencies from 5 Hz to 1 MHz, and an applied voltage amplitude of 0.1 V. Li^+-blocking Au film electrodes were formed on both parallel surfaces of the pellet by sputtering for the conductivity measurement.

2.2. Evaluation of Stability for LLZT/Li Interface

To evaluate interface stability between LLZT and Li metal, we constructed the Li/LLZT/Li symmetric cell with a cell fixture (Figure 1) in an Ar-filled grove box by attaching Li metal foils on the polished top and bottom surfaces of an LLZT pellet with a thickness between 1.90 and 1.95 mm. Heat treatments at different temperatures (100–175 °C) and time (1–5 h) were applied to the cell after cell construction. Interfacial charge-transfer resistance $R_{\text{Li-LLZT}}$ between LLZT and Li was evaluated by A/C impedance measurements. The influence of $R_{\text{Li-LLZT}}$ on the stability of the Li deposition and dissolution reaction was also investigated using galvanostatic cycling testing of a symmetric cell at 25 °C using a Battery Test System (TOSCAT-3100, TOYO SYSTEM, Iwaki, Japan). The microstructure of the LLZT pellet after the galvanostatic testing was observed by SEM (SU8000 Type II, HITACHI, Tokyo, Japan).

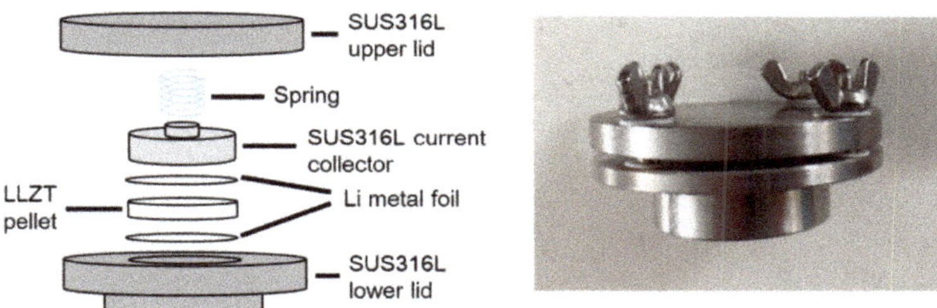

Figure 1. Illustration (left) and photo of cell fixture for composing a lithium-garnet-type $Li_{7-x}La_3Zr_{2-x}Ta_xO_{12}$-lithium (Li/LLZT/Li) symmetric cell.

3. Results and Discussion

3.1. Characterization of Sintered LLZT Pellets

The XRD pattern and SEM image of the fractured surface of sintered LLZT are shown in Figure 2. The calculated pattern with the reported structural parameters for cubic LLZ is also plotted in Figure 2a [7]. All the peaks for the LLZT sample were well indexed as cubic garnet-type structures with a space group $Ia\bar{3}d$; no other secondary phases were observed. Compared with the calculated pattern for cubic LLZ, all peaks of LLZT shifted toward a higher angle 2θ. This was attributed to the reduction in the lattice size by substitution of Zr^{4+} (72 pm) by smaller Ta^{5+} (64 pm) [15–23]. As shown in Figure 2b, the average grain size in the sintered LLZT was around 5 μm and all grains were in good contact with each other to form a dense structure. The density of the pellet was determined from their weight and physical dimensions. The relative density (measured density normalized by the theoretical density) of LLZT used in this work was around 92–93%.

The ionic conductivity of LLZT was examined by A/C impedance spectroscopy using a Li^+ ion blocking silver (Au) electrode. Figure 3a shows the Nyquist plots of A/C impedance measured at 27, 50, and 75 °C for LLZT. A part of the small semicircle and linear portion data were obtained in high and low frequency regions, indicating that the conduction is primarily ionic in nature. The intercept point of the linear tail in the low frequency range with a real axis corresponds to the sum of the bulk and grain-boundary resistances. Total conductivity σ was calculated by the inverse of the total (bulk and grain boundary) resistance and geometrical parameters of the pellet, which was estimated to be 0.93 mS cm^{-1} at 27 °C. Notably, we prepared many LLZT pellets for stability testing described later, and all pellets reproducibly showed σ = 0.9–1.0 mS cm^{-1} at 27 °C.

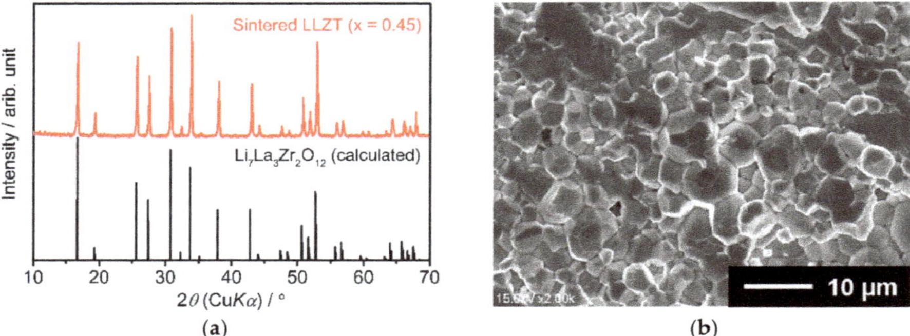

(a) (b)

Figure 2. (**a**) X-ray diffraction (XRD) pattern and (**b**) scanning electron microscope (SEM) image of fractured cross section for sintered LLZT.

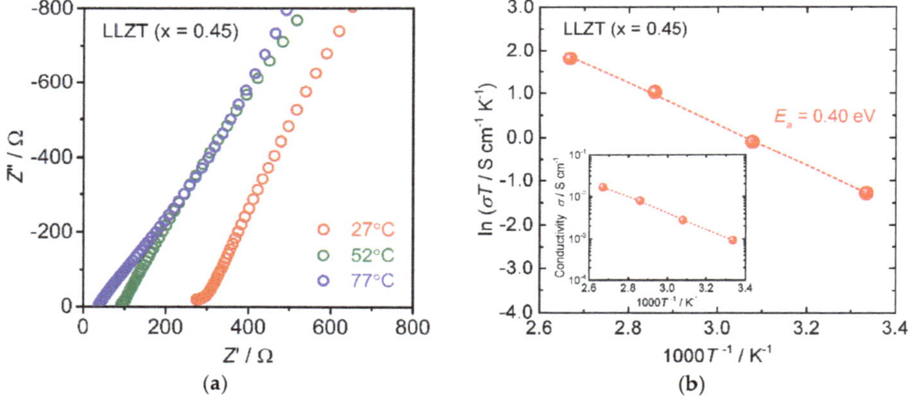

(a) (b)

Figure 3. (**a**) Nyquist plot of alternating current (A/C) impedance of LLZT at 27 °C. (**b**) Arrhenius plot of the conductivity σ for LLZT plotted against the inverse of measurement temperature.

Figure 3b shows the variation in σ for an LLZT pellet as a function of an inverse of temperature 1000 T^{-1}. σ increased monotonically as the temperature increased and reached 16 mS cm^{-1} at 100 °C. In addition, the temperature dependence of σ was well-expressed by the Arrhenius equation as $\sigma = \sigma_0 \exp(-Ea/(k_BT))$, where σ_0 is constant, E_a is the activation energy of conductivity, and k_B is the Boltzmann constant (1.381 × 10^{-23} J K^{-1}). From the slope data of σT in Figure 3b, E_a of LLZT was calculated as 0.40 eV, which is nearly the same as the Al-free LLZT reported in the literature [19,21,22].

3.2. Interfacial Charge Transfer Resistance between LLZT and Li Metal Electrode

To examine the effect of a heat treatment condition on $R_{Li\text{-}LLZT}$ between LLZT and Li, we measured the A/C impedance of Li/LLZT/Li symmetric cells with various heat treatment conditions. Figure 4a shows the Nyquist plots of A/C impedance at 27 °C for Li/LLZT/Li cell both before and after heat treatment at different temperatures for one hour. The data in Figure 4a were obtained in a continuous experiment with one specific cell. Firstly, we measured the impedance for the cell before applying heat treatment, then the cell was heat-treated at 100 °C for one hour, cooled down to 27 °C, and the impedance measurements were recorded. After that, measurements for the cell heat-treated at 150 or 175 °C for one hour were recorded in sequence.

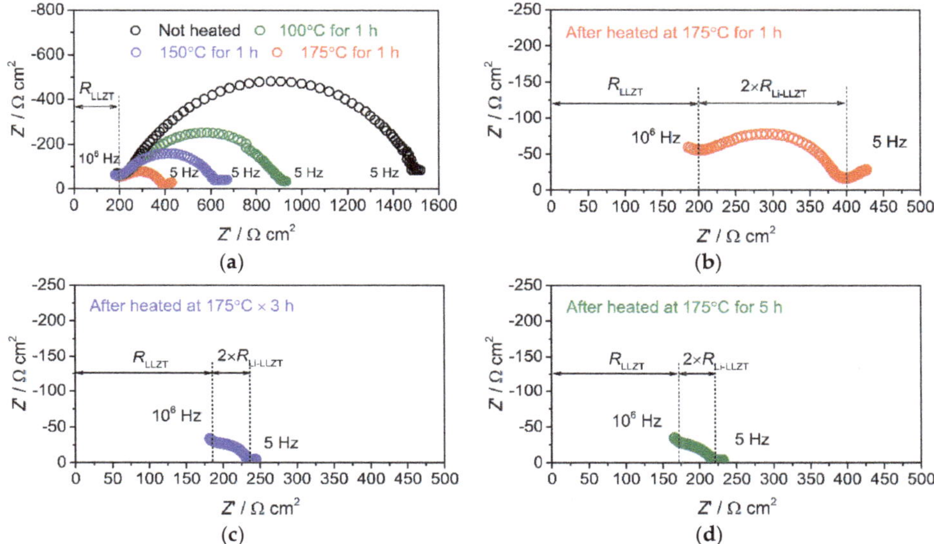

Figure 4. (**a**) Nyquist plots of AC impedance for Li/LLZT/Li symmetric cell at 27 °C before and after heat treatment at 100, 150, and 175 °C for one hour. Different Li/LLZT/ Li cells heat treated at 175 °C for (**b**) one, (**c**) three, and (**d**) five hours.

In the frequency range of 5 Hz to 10^5 Hz, one semi-circle was clearly visible in the Li/LLZT/Li cell before and after applying heat treatment at different temperatures. This semi-circle was not observed in the Au/LLZT/Au cell shown in Figure 3a. At the high frequency range from 10^5 to 10^6 Hz, the plots almost overlap, suggesting that the data above 10^5 Hz are mainly attributed to the ionic conducting property of LLZT pellet. The data below 10^5 Hz correspond to the charge-transfer characteristics at the interface between LLZT and the two Li metal electrodes, and the diameter of the semicircle below 10^5 Hz corresponds to $2 \times R_{\text{Li-LLZT}}$. As shown in Figure 4a, $R_{\text{Li-LLZT}}$ decreased gradually as the heat treatment temperature increased. Before applying heat treatment, $R_{\text{Li-LLZT}}$ at 27 °C was 650 Ω cm^2 and decreased to 350 Ω cm^2, 200 Ω cm^2, and 100 Ω cm^2, after applying heat treatment at 100, 150, and 175 °C for one hour, respectively.

For further reduction of $R_{\text{Li-LLZT}}$, we investigated the influence of heat treatment time at 175 °C, which is only 5 °C lower than the melting point of Li. Nyquist plots of A/C impedance at 27 °C for Li/LLZT/Li cells heat-treated at 175 °C for one three, and five hours are shown in Figure 4b–d. $R_{\text{Li-LLZT}}$ was reduced further by increasing the heat treatment time, but both the cells heat treated at 175 °C for three and five hours showed the same $R_{\text{Li-LLZT}}$ of 25 Ω cm^2 at 27 °C, indicating that the effect of heat treatment time on the reduction of $R_{\text{Li-LLZT}}$ was nearly saturated at three hours. We think that the reduction in $R_{\text{Li-LLZT}}$ obtained by optimizing the heat treatment condition resulted from improved wetting and contact between LLZT and Li.

Using the cell heat-treated at 175 °C for five hours, the temperature dependence of $R_{\text{Li-LLZT}}$ was investigated and the results are shown in Figure 5. The measurements were recorded in both heating and cooling processes at 27–100 °C. As shown in Figure 5a,b, the impedance data measured at a fixed temperature in heating and cooling processes were nearly identical, suggesting that the resistances of LLZT, R_{LLZT}, and $R_{Li\text{-}LLZT}$ were hardly affected by temperatures below 100 °C.

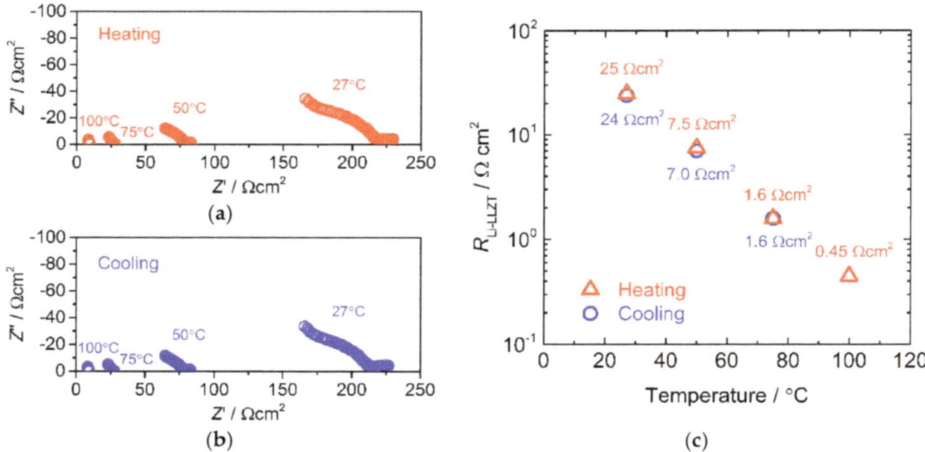

Figure 5. A/C impedance measured at different temperatures for Li/LLZT/Li symmetric cells after heat treatment at 175 °C for five hours: (**a**) during heating process from 27 to 100 °C and (**b**) cooling process from 100 to 27 °C. (**c**) Interfacial charge transfer resistance $R_{Li\text{-}LLZT}$ as a function of temperature.

3.3. Stability against Li Deposition and Dissolution Reaction at the Interface

Next, we discuss the stability against Li deposition and dissolution reaction at the interface between LLZT and Li. Figure 6 shows the results for galvanostatic cycling at 25 °C in Li/LLZT/Li symmetric cells with different $R_{Li\text{-}LLZT}$ (27 °C) of 100 and 25 Ω cm^2. In this work, "one cycle" is defined as the process of passing a current in both directions, and then resting for 600 s during each cycle. Cell voltage was determined as the product of current and cell resistance (= R_{LLZT} + $R_{Li\text{-}LLZT}$). In both cells, Ohmic behavior was observed at low current densities followed by deviation from Ohmic behavior at high current densities. Similar behavior has been confirmed in the literature [33,34,40–46]. Furthermore, the polarization was not symmetric against the current direction, which could have been caused by an irreversibility in Li deposition and dissolution reaction at each interface between LLZT and Li in a symmetric cell [43,46]. The cell with lower $R_{Li\text{-}LLZT}$ (Figure 6b) shows lesser polarization than the cell with higher $R_{Li\text{-}LLZT}$ (Figure 6a) and maintained Ohmic behavior under higher current densities around 0.1 mA cm^{-2}. The cell voltage suddenly dropped to 0 V at 0.11 mA cm^{-2} for the cell with an $R_{Li\text{-}LLZT}$ of 100 Ω cm^2, whereas the voltage drop in the cell with a lower $R_{Li\text{-}LLZT}$ of 25 Ω cm^2 occurred at 0.36 mA cm^{-2}. This indicates that the short circuit occurred inside the cell and the reduction of $R_{Li\text{-}LLZT}$ had a positive effect on enhancing the tolerance against the short circuit. Reducing $R_{Li\text{-}LLZT}$ likely resulted in more uniform current density, thus the tolerance against the short circuit improved. We also performed galvanostatic cycling for the symmetric cell at 100 °C. Increasing temperature improved wetting and contact between Li and LLZT. The $R_{Li\text{-}LLZT}$ of the tested cell was 25 Ω cm^2 at 27 °C and decreased to 0.5 Ω cm^2 at 100 °C. As expected, the current density at which the short circuit occurred increased to 1.2 mA cm^{-2} (Figure S1).

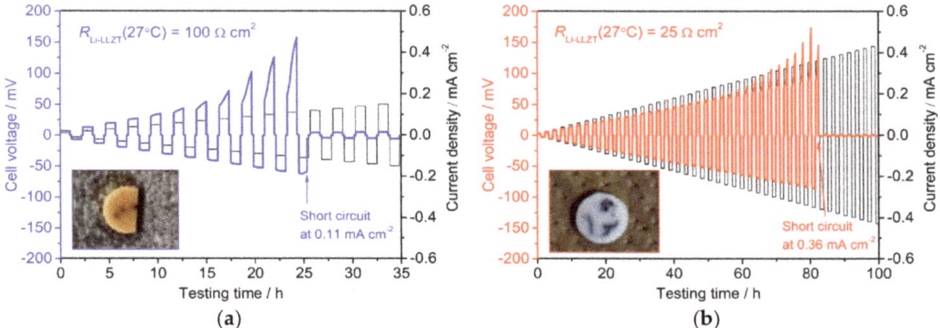

Figure 6. Galvanostatic cycling at 25 °C for Li/LLZT/Li cells with different $R_{\text{Li-LLZT}}$: (**a**) $R_{\text{Li-LLZT}}$ (27 °C) = 100 Ω cm^2 and (**b**) $R_{\text{Li-LLZT}}$ (27 °C) = 25 Ω cm^2. The inset in each graph is a photo of an LLZT pellet removed from each cell after a short circuit occurred.

The photos of the LLZT pellets removed from both cells after short circuits occurred are also shown in the insets of both graphs. Some black spots and cracks were clearly confirmed on the pellet surface. Moreover, some of the cracks penetrated along a thickness direction and reached the opposing pellet surface, indicating that the short circuit locally occurred around these cracks due to the propagation of Li dendrite into SE [33,47–49]. We also observed the microstructures of fractured cross sections of LLZT pellets using SEM and the results are shown in Figure 7. SEM observation was performed around the black spot. In Figure 7a, many fiber-like materials with submicron-sized diameters were confirmed between the LLZT grains in the pellet removed from the cell with $R_{\text{Li-LLZT}}$ = 100 Ω cm^2. However, in Figure 7b, a whisker-like material two to three μm in diameter was confirmed to propagate into the pellet removed from the cell with $R_{\text{Li-LLZT}}$ = 25 Ω cm^2. We think that these materials observed in LLZT pellets are Li dendrites that cause the short circuit during galvanostatic cycling test depicted in Figure 5.

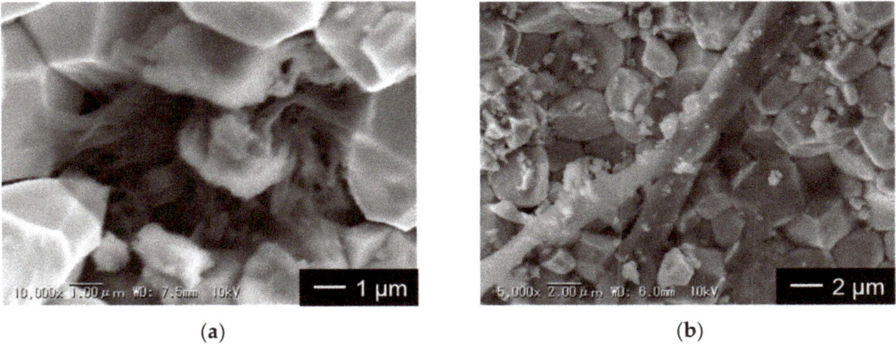

Figure 7. SEM images for cross sections of LLZT pellets removed from Li/LLZT/Li cells after a short circuit occurred: (**a**) $R_{\text{Li-LLZT}}$ (27 °C) = 100 Ω cm^2 and (**b**) $R_{\text{Li-LLZT}}$ (27 °C) = 25 Ω cm^2.

We fabricated other symmetric cells with $R_{\text{Li-LLZT}}$ = 25 Ω cm^2 and their cycle stabilities were investigated at 25 °C and at current densities of 0.15 and 0.30 mA cm^{-2} for application times of one and two hours. The results are summarized in Figure 8. Notably, the measurements at these four different conditions were continuously performed in one specific cell. The cell was stably tested over 100 cycles without the occurrence of a short circuit at the conditions of 0.15 mA cm^{-2} for one hour,

0.15 mA cm^{-2} for two hours, and 0.30 mA cm^{-2} for one hour, whereas the polarization and deviation from Ohmic low become remarkable with increasing magnitude of current density and its application time. However, at 0.30 mA cm^{-2} for two hours, cell voltage increased abnormally in the first half cycle, and then a short circuit occurred during the second cycle. Under this condition, the cell could not be cycled stably.

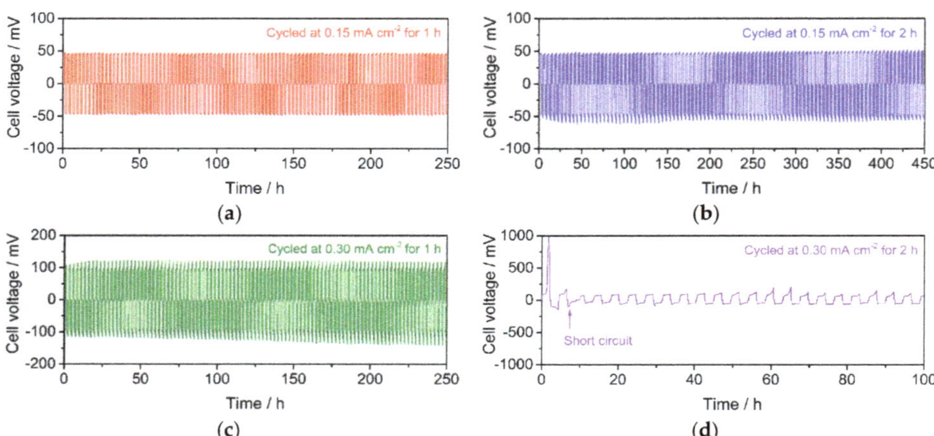

Figure 8. Galvanostatic cycling at 25 °C for Li/LLZT/Li cell with $R_{Li\text{-}LLZT}$ (27 °C) = 25 Ω cm^2: (**a**) cycled at 0.15 mA cm^{-2} for one hour, (**b**) cycled at 0.15 mA cm^{-2} for two hours, (**c**) cycled at 0.30 mA cm^{-2} for one hour, and (**d**) cycled at 0.30 mA cm^{-2} for two hours. The measurements were continuously recorded from (**a**) to (**d**).

Before and after each galvanostatic cycling under different conditions, we also checked the cell impedance at room temperature. The measurement results are shown in Figure 9. Before galvanostatic cycling, the cell resistance was around 230 Ω cm^2. After being cycled at 0.15 mA cm^{-2} for one hour, 0.15 mA cm^{-2} for two hours, and 0.30 mA cm^{-2} for one hour, the cell demonstrated similar results and cell resistance ranging from 215 to 225 Ω cm^2. This is consistent with the results for the stable galvanostatic cycling shown in Figure 8a–c. However, after being cycled at 0.30 mA cm^{-2} for two hours, the plot significantly deviated from the others and the cell resistance decreased to 65 Ω cm^2.

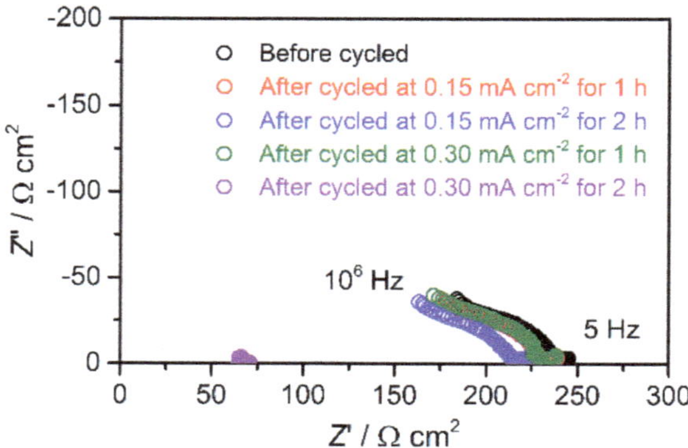

Figure 9. A/C impedance measured at 27 °C for Li/LLZT/Li cell before and after galvanostatic cycling with different conditions described in the caption of Figure 8.

The conditions for the stable cycling of a symmetric cell were expected to be influenced not only by the magnitude of the current density but also by the capacity for Li deposition at the interface between LLZT and Li. Using the data for the first half cycle in Figure 8c,d, the change in cell voltage at 25 °C and 0.30 mA cm^{-2} are plotted as a function of area specific capacity for Li deposition (Figure 10). The plots of the cell voltage tested at 0.30 mA cm^{-2} for one and two hours nearly overlapped at the capacity for Li deposition below 0.30 mAh cm^{-2} and an abnormal increase in cell voltage was caused at a capacity above 0.35 mAh cm^{-2}. Similar behavior was confirmed when the cell was cycled at smaller current density of 0.15 mA cm^{-2}. As shown in Figure 8a,b, the cell was stably cycled at 0.15 mA cm^{-2} for one and two hours, but a short circuit occurred at 0.15 mA cm^{-2} for three hours (Figure S2). The abnormal increase in cell voltage was also confirmed at the capacity for Li deposition above 0.30 mAh cm^{-2} (Figure S3). Although our measurement conditions in this work were limited, the threshold of the area for specific Li deposition capacity for stable was roughly expected to be around 0.30 mAh cm^{-2}. When we used the LLZT prepared in this work, the $R_{\text{Li-LLZT}}$ at room temperature decreased to 20–30 Ω cm^2, and the current density for galvanostatic cycling was set below 0.30 mA cm^{-2}.

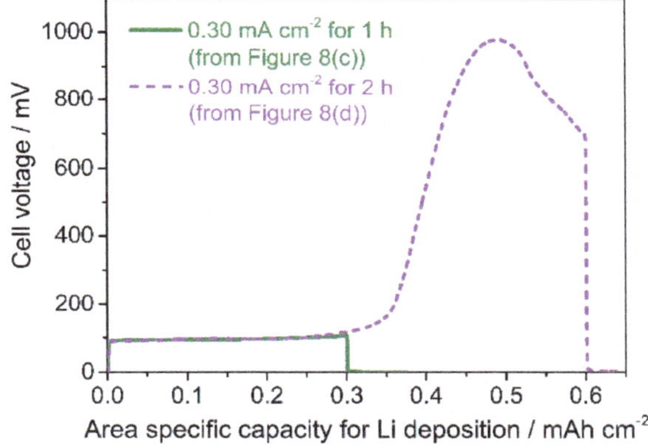

Figure 10. Cell voltage at 25 °C and 0.30 mA cm^{-2} for one or two hours in Li/LLZT/Li cell with $R_{\text{Li-LLZT}}$ (27 °C) = 25 Ω cm^2, as a function of area specific capacity for Li deposition. The plotted data were calculated from the data from the first cycle in Figure 8c,d.

Together with further examination of the propagation mechanism of Li dendrite into garnet-type SE, both the critical current density at which the short circuit occurs and the threshold in the capacity for Li deposition for stable cycling must be enhanced further to realize a solid-state battery with a garnet-type oxide SE and Li metal anode. We believe that this could be achieved by reducing the interfacial charge-transfer resistance further and by structural improvement of the garnet-type SE by decreasing the porosity [34,47,48], controlling grain size [41,44], and modifying the grain boundary [45,50].

4. Conclusions

We investigated the influence of heating temperature and time on the interfacial resistance between LLZT and a Li metal electrode using a Li/LLZT/Li symmetric cell. In addition, the effect of the interfacial charge-transfer resistance $R_{\text{Li-LLZT}}$ on the stability of Li deposition and dissolution was investigated using a galvanostatic cycling test. The lowest $R_{\text{Li-LLZT}}$ of 25 Ω cm^2 at room temperature was obtained by heating at 175 °C for three to five hours after contacting LLZT with Li. We confirmed that reducing interfacial resistance is effective for enhancing the current density at which the short circuit occurs by propagating the Li dendrite into LLZT. The conditions for stable cycling of a symmetric cell are influenced not only by the magnitude of the current density but also by the capacity for Li deposition at the interface between LLZT and Li.

Supplementary Materials: The following are available online at http://www.mdpi.com/2313-0105/4/2/26/s1, Figure S1: Galvanostatic cycling at 100 °C for Li/LLZT/Li cell with $R_{Li\text{-}LLZT}$ = 25 cm^2 at 27 °C and 0.5 cm^2 at 100 °C, Figure S2: Galvanostatic cycling at 25 °C for Li/LLZT/Li cell with $R_{Li\text{-}LLZT}$ (27 °C) = 28 cm^2 tested at 0.15 mA cm^{-2} for 3 h, Figure S3: Cell voltage at 25 °C and 0.15 mA cm^{-2} for 1, 2 and 3 h in Li/LLZT/Li cells as a function of area specific capacity for Li deposition.

Author Contributions: R.I., S.Y., H.H., M.S. conceived, designed, and executed the experiments. All authors were involved in the analysis of the results. R.I. wrote the manuscript. All authors were involved in the discussion of the results and manuscript.

Acknowledgments: This work was partly supported by JSPS KAKENHI grant numbers 16K06218 and 16KK0127 from the Japan Society for the Promotion of Science (JSPS) and Research Foundation for the Electro-technology of Chubu (R-28241).

Conflicts of Interest: The authors declare no conflict of interest.

References

1. Armand, M.; Tarascon, J.M. Building better batteries. *Nature* **2008**, *451*, 652–657. [CrossRef] [PubMed]
2. Scrosati, B.; Garche, J. Lithium batteries: Status, prospects and future. *J. Power Sources* **2010**, *195*, 2419–2430. [CrossRef]
3. Goodenough, J.B.; Kim, Y. Challenges for rechargeable batteries. *J. Power Sources* **2011**, *196*, 6688–6694. [CrossRef]
4. Fergus, J.W. Ceramic and polymeric solid electrolytes for lithium-ion batteries. *J. Power Sources* **2010**, *195*, 4554–4569. [CrossRef]
5. Takada, K. Progress and prospective of solid-state lithium batteries. *Acta Mater.* **2012**, *61*, 759–770. [CrossRef]
6. Tatsumisago, M.; Nagao, M.; Hayashi, A. Recent development of sulfide solid electrolytes and interfacial modification for all-solid-state rechargeable lithium batteries. *J. Asian Ceram. Soc.* **2013**, *1*, 117–125. [CrossRef]
7. Knauth, P. Inorganic solid Li ion conductors: An overview. *Solid State Ion.* **2009**, *180*, 911–916. [CrossRef]
8. Ren, Y.; Chen, K.; Chen, R.; Liu, T.; Zhang, Y.; Nan, C.-W. Oxide electrolytes for lithium batteries. *J. Am. Ceram. Soc.* **2015**, *98*, 3603–3623. [CrossRef]
9. Murugan, R.; Thangadurai, V.; Weppner, W. Fast lithium ion conduction in garnet-type $Li_7La_3Zr_2O_{12}$. *Angew. Chem. Int. Ed.* **2007**, *46*, 7778–7781. [CrossRef] [PubMed]
10. Awaka, J.; Takashima, A.; Kataoka, K.; Kijima, N.; Idemoto, Y.; Akimoto, J. Crystal structure of fast lithium-ion-conducting cubic $Li_7La_3Zr_2O_{12}$. *Chem. Lett.* **2011**, *40*, 60–62. [CrossRef]
11. Awaka, J.; Kijima, N.; Hayakawa, H.; Akimoto, J. Synthesis and structure analysis of tetragonal $Li_7La_3Zr_2O_{12}$ with the garnet-related type structure. *J. Solid State Chem.* **2009**, *182*, 2046–2052. [CrossRef]
12. Geiger, C.A.; Alekseev, E.; Lazic, B.; Fisch, M.; Armbruster, T.; Langner, R.; Fechtelkord, M.; Kim, N.; Pettke, T.; Weppner, W. Crystal chemistry and stability of "$Li_7La_3Zr_2O_{12}$" garnet: A fast lithium-ion conductor. *Inorg. Chem.* **2011**, *50*, 1089–1097. [CrossRef] [PubMed]
13. Ohta, S.; Kobayashi, T.; Asaoka, T. High lithium ionic conductivity in the garnet-type oxide $Li_{7-X}La_3(Zr_{2-X},Nb_X)O_{12}$ (X = 0–2). *J. Power Sources* **2011**, *196*, 3342–3345. [CrossRef]
14. Kihira, Y.; Ohta, S.; Imagawa, H.; Asaoka, T. Effect of simultaneous substitution of alkali earth metals and Nb in $Li_7La_3Zr_2O_{12}$ on lithium-ion conductivity. *ECS Electrochem. Lett.* **2013**, *2*, A56–A59. [CrossRef]
15. Li, Y.; Han, J.-T.; Wang, C.-A.; Xie, H.; Goodenough, J.B. Optimizing Li^+ conductivity in a garnet framework. *J. Mater. Chem.* **2012**, *22*, 15357–15361. [CrossRef]
16. Buschmann, H.; Berendts, S.; Mogwitz, B.; Janek, J. Lithium metal electrode kinetics and ionic conductivity of the solid lithium ion conductors "$Li_7La_3Zr_2O_{12}$" and $Li_{7-x}La_3Zr_{2-x}Ta_xO_{12}$ with garnet-type structure. *J. Power Sources* **2012**, *206*, 236–244. [CrossRef]
17. Logéat, A.; Köhler, T.; Eisele, U.; Stiaszny, B.; Harzer, A.; Tovar, M.; Senyshyn, A.; Ehrenberg, H.; Kozinsky, B. From order to disorder: The structure of lithium-conducting garnets $Li_{7-x}La_3Ta_xZr_{2-x}O_{12}$ (x = 0–2). *Solid State Ion.* **2012**, *206*, 33–38. [CrossRef]
18. Wang, Y.; Wei, L. High ionic conductivity lithium garnet oxides of $Li_{7-x}La_3Zr_{2-x}Ta_xO_{12}$ compositions. *Electrochem. Solid State Lett.* **2012**, *15*, A68–A71. [CrossRef]

19. Thompson, T.; Sharafi, A.; Johannes, M.D.; Huq, A.; Allen, J.L.; Wolfenstine, J.; Sakamoto, J. Tale of two sites: On defining the carrier concentration in garnet-based ionic conductors for advanced Li batteries. *Adv. Energy Mater.* **2015**, *5*, 1500096. [CrossRef]
20. Thompson, T.; Wolfenstine, J.; Allen, J.L.; Johannes, M.; Huq, A.; Davida, I.N.; Sakamoto, J. Tetragonal vs. cubic phase stability in Al-free Ta doped $Li_7La_3Zr_2O_{12}$ (LLZO). *J. Mater. Chem. A* **2014**, *2*, 13431–13436. [CrossRef]
21. Inada, R.; Kusakabe, K.; Tanaka, T.; Kudo, S.; Sakurai, Y. Synthesis and properties of Al-free $Li_{7-x}La_3Zr_{2-x}Ta_xO_{12}$ garnet related oxides. *Solid State Ion.* **2014**, *262*, 568–572. [CrossRef]
22. Inada, R.; Yasuda, S.; Tojo, T.; Sakurai, Y. Development of lithium-stuffed garnet-type oxide solid electrolytes with high ionic conductivity for application to all-solid-state batteries. *Front. Energy Res.* **2016**, *4*, 28. [CrossRef]
23. Ren, Y.; Deng, H.; Chen, R.; Shen, Y.; Lin, Y.; Nan, C.-W. Effects of Li source on microstructure and ionic conductivity of Al-contained $Li_{6.75}La_3Zr_{1.75}Ta_{0.25}O_{12}$ ceramics. *J. Eur. Ceram. Soc.* **2015**, *35*, 561–572. [CrossRef]
24. Dhivya, L.; Janani, N.; Palanivel, B.; Murugan, R. Li^+ transport properties of W substituted $Li_7La_3Zr_2O_{12}$ cubic lithium garnets. *AIP Adv.* **2013**, *3*, 082115. [CrossRef]
25. Li, Y.; Wang, Z.; Cao, Y.; Du, F.; Chen, C.; Cui, Z.; Guo, X. W-doped $Li_7La_3Zr_2O_{12}$ ceramic electrolytes for solid state Li-ion batteries. *Electrochim. Acta* **2015**, *180*, 37–42. [CrossRef]
26. Bottke, P.; Rettenwander, D.; Schmidt, W.; Amthauer, G.; Wilkening, M. Ion dynamics in solid electrolytes: NMR reveals the elementary steps of Li^+ hopping in the garnet $Li_{6.5}La_3Zr_{1.75}Mo_{0.25}O_{12}$. *Chem. Mater.* **2015**, *27*, 6571–6582. [CrossRef]
27. Ohta, S.; Komagata, S.; Seki, J.; Saeki, T.; Morishita, S.; Asaoka, T. All solid-state lithium ion battery using garnet-type oxide and Li_3BO_3 solid electrolytes fabricated by screen-printing. *J. Power Sources* **2013**, *238*, 53–56. [CrossRef]
28. Ohta, S.; Seki, J.; Yagi, Y.; Kihira, Y.; Tani, T.; Asaoka, T. Co-sinterable lithium garnet-type oxide electrolyte with cathode for all-solid-state lithium ion battery. *J. Power Sources* **2014**, *265*, 40–44. [CrossRef]
29. Nemori, H.; Matsuda, Y.; Mitsuoka, S.; Matsui, M.; Yamamoto, O.; Takeda, Y.; Imanishi, N. Stability of garnet-type solid electrolyte $Li_xLa_3A_{2-y}B_yO_{12}$ (A = Nb or Ta, B = Sc or Zr). *Solid State Ion.* **2015**, *282*, 7–12. [CrossRef]
30. Kim, Y.; Yoo, A.; Schmidt, R.; Sharafi, A.; Lee, H.; Wolfenstine, J.; Sakamoto, J. Electrochemical Stability of $Li_{6.5}La_3Zr_{1.5}M_{0.5}O_{12}$ (M = Nb or Ta) against Metallic Lithium. *Front. Energy Res.* **2016**, *4*, 20. [CrossRef]
31. Xie, H.; Park, K.-S.; Song, J.; Goodenough, J.B. Reversible lithium insertion in the garnet framework of $Li_3Nd_3W_2O_{12}$. *Electrochem. Commun.* **2012**, *19*, 135–137. [CrossRef]
32. Hofstetter, K.; Samson, A.J.; Narayanan, S.; Thangadurai, V. Present understanding of the stability of Li-stuffed garnets with moisture, carbon dioxide, and metallic lithium. *J. Power Sources* **2018**, *390*, 297–312. [CrossRef]
33. Sharafi, A.; Meyer, H.M.; Nanda, J.; Wolfenstine, J.; Sakamoto, J. Characterizing the Li–$Li_7La_3Zr_2O_{12}$ interface stability and kinetics as a function of temperature and current density. *J. Power Sources* **2016**, *302*, 135–139. [CrossRef]
34. Basappa, R.H.; Ito, T.; Yamada, H. Contact between garnet-type solid electrolyte and lithium metal anode: Influence on charge transfer resistance and short circuit prevention. *J. Electrochem. Soc.* **2017**, *164*, A666–A671. [CrossRef]
35. Tsai, C.-L.; Roddatis, V.; Chandran, C.V.; Ma, Q.; Uhlenbruck, S.; Bram, M.; Heitjans, P.; Guillon, O. $Li_7La_3Zr_2O_{12}$ interface modification for Li dendrite prevention. *ACS Appl. Mater. Interfaces* **2016**, *8*, 10617–10626. [CrossRef] [PubMed]
36. Luo, W.; Gong, Y.; Zhu, Y.; Fu, K.K.; Dai, J.; Lacey, S.D.; Wang, C.; Liu, B.; Han, X.; Mo, Y.; et al. Transition from superlithiophobicity to superlithiophilicity of garnet solid-state electrolyte. *J. Am. Chem. Soc.* **2016**, *138*, 12258–12262. [CrossRef] [PubMed]
37. Luo, W.; Gong, Y.; Zhu, Y.; Li, Y.; Yao, Y.; Zhang, Y.; Fu, K.; Pastel, G.; Lin, C.-F.; Mo, Y.; et al. Reducing interfacial resistance between garnet-structured solid-state electrolyte and Li-metal Anode by a germanium layer. *Adv. Mater.* **2017**, *29*, 1606042. [CrossRef] [PubMed]
38. Han, X.; Gong, Y.; Fu, K.; He, X.; Hitz, G.T.; Dai, J.; Pearse, A.; Liu, B.; Wang, H.; Rublo, G.; et al. Negating interfacial impedance in garnet-based solid-state Li metal batteries. *Nat. Mater.* **2017**, *16*, 572–579. [CrossRef] [PubMed]

39. Wang, C.; Gong, Y.; Liu, B.; Fu, K.; Yao, Y.; Hitz, E.; Li, Y.; Dai, J.; Xu, S.; Luo, W.; et al. Conformal, nanoscale ZnO surface modification of garnet based solid-state electrolyte for lithium metal anodes. *Nano Lett.* **2017**, *17*, 565–571. [CrossRef] [PubMed]
40. Sharafi, A.; Kazyak, E.; Davis, A.L.; Yu, S.; Thompson, T.; Siegel, D.J.; Dasgupta, N.P.; Sakamoto, J. Surface chemistry mechanism of ultra-low interfacial resistance in the solid-state electrolyte $Li_7La_3Zr_2O_{12}$. *Chem. Mater.* **2017**, *29*, 7961–7968. [CrossRef]
41. Sharafi, A.; Meyer, H.M.; Nanda, J.; Wolfenstine, J.; Sakamoto, J. Controlling and correlating the effect of grain size with the mechanical and electrochemical properties of $Li_7La_3Zr_2O_{12}$ solid-state electrolyte. *J. Mater. Chem. A* **2017**, *5*, 21491–21504. [CrossRef]
42. Wang, M.; Sakamoto, J. Correlating the interface resistance and surface adhesion of the Li metal solid electrolyte interface. *J. Power Sources* **2018**, *377*, 7–11. [CrossRef]
43. Yonemoto, F.; Nishimura, A.; Motoyama, M.; Tsuchimine, N.; Kobayashi, S.; Iriyama, Y. Temperature effects on cycling stability of Li plating/stripping on Ta-doped $Li_7La_3Zr_2O_{12}$. *J. Power Sources* **2017**, *343*, 207–215. [CrossRef]
44. Yamada, H.; Ito, T.; Rajendra, H.B. Sintering mechanisms of high-performance garnet-type solid electrolyte densified by spark plasma sintering. *Electrochim. Acta* **2016**, *222*, 648–656. [CrossRef]
45. Basappa, R.H.; Ito, T.; Morimura, T.; Bekarevich, R.; Mitsuishi, K.; Yamada, H. Grain boundary modification to suppress lithium penetration through garnet-type solid electrolyte. *J. Power Sources* **2017**, *363*, 145–152. [CrossRef]
46. Koshikawa, H.; Matsuda, S.; Kamiya, K.; Miyayama, M.; Kubo, Y.; Uosaki, K.; Hashimoto, K.; Nakanishi, S. Dynamic changes in charge-transfer resistance at Li metal/$Li_7La_3Zr_2O_{12}$ interfaces during electrochemical Li dissolution/deposition cycles. *J. Power Sources* **2018**, *376*, 147–151. [CrossRef]
47. Ren, Y.; Shen, Y.; Lin, Y.; Nan, C.-W. Direct observation of lithium dendrites inside garnet-type lithium-ion solid electrolyte. *Electrochem. Commun.* **2015**, *57*, 27–30. [CrossRef]
48. Suzuki, Y.; Kami, K.; Watanabe, K.; Watanabe, A.; Saito, N.; Ohnishi, T.; Takada, K.; Sudo, R.; Imanishi, N. Transparent cubic garnet-type solid electrolyte of Al_2O_3-doped $Li_7La_3Zr_2O_{12}$. *Solid State Ion.* **2015**, *278*, 172–176. [CrossRef]
49. Cheng, E.J.; Sharafi, A.; Sakamoto, J. Intergranular Li metal propagation through polycrystalline $Li_{6.25}Al_{0.25}La_3Zr_2O_{12}$ ceramic electrolyte. *Electrochim. Acta* **2017**, *223*, 85–91. [CrossRef]
50. Xu, B.; Li, W.; Duan, H.; Wang, H.; Guo, Y.; Li, H.; Liu, H. Li_3PO_4-added garnet-type $Li_{6.5}La_3Zr_{1.5}Ta_{0.5}O_{12}$ for Li-dendrite suppression. *J. Power Sources* **2017**, *354*, 68–73. [CrossRef]

Article

High-Performance $Na_{0.44}MnO_2$ Slabs for Sodium-Ion Batteries Obtained through Urea-Based Solution Combustion Synthesis

Chiara Ferrara [1], Cristina Tealdi [1,*], Valentina Dall'Asta [1], Daniel Buchholz [2,3], Luciana G. Chagas [2,3], Eliana Quartarone [1], Vittorio Berbenni [1] and Stefano Passerini [2,3]

1 Department of Chemistry and INSTM, University of Pavia, Via Taramelli 12, 27100 Pavia, Italy; chiara.ferrara01@universitadipavia.it (C.F.); vale.dallasta@gmail.com (V.D.); eliana.quartarone@unipv.it (E.Q.); vittorio.berbenni@unipv.it (V.B.)

2 Helmholtz Institute Ulm (HIU), Helmholtzstraße 11, 89081 Ulm, Germany; daniel.buchholz@kit.edu (D.B.); luciana.chagas@kit.edu (L.G.C.); stefano.passerini@kit.edu (S.P.)

3 Karlsruhe Institute of Technology (KIT), P.O. Box 3640, 76021 Karlsruhe, Germany

* Correspondence: cristina.tealdi@unipv.it; Tel.: +39-382-98-7569

Received: 30 December 2017; Accepted: 31 January 2018; Published: 9 February 2018

Abstract: One of the primary targets of current research in the field of energy storage and conversion is the identification of easy, low-cost approaches for synthesizing cell active materials. Herein, we present a novel method for preparing nanometric slabs of $Na_{0.44}MnO_2$, making use of the eco-friendly urea within a solution synthesis approach. This kind of preparation greatly reduces the time of reaction, decreases the thermal treatment temperature, and allows the obtaining of particles with smaller dimensions compared with those obtained through conventional solid-state synthesis. Such a decrease in particle size guarantees improved electrochemical performance, particularly at high current densities, where kinetic limitations become relevant. Indeed, the materials produced via solution synthesis outperform those prepared via solid-state synthesis both at 2 C, (95 mA h g^{-1} vs. 85 mA h g^{-1}, respectively) and 5 C, (78 mA h g^{-1} vs. 68.5 mA h g^{-1}, respectively). Additionally, the former material is rather stable over 200 cycles, with a high capacity retention of 75.7%.

Keywords: sodium-ion battery; cathode; solution combustion synthesis; capacity retention; $Na_{0.44}MnO_2$

1. Introduction

Energy storage is a key challenge of the present time. Academic and industrial research is focusing on the development of optimized materials and fabrication processes in order to meet stringent requirements in terms of electrochemical characteristics, while concomitantly achieving the preparation and commercialization of sustainable products, both from an economic and an environmental perspective [1]. For this reason, attention is being devoted to the search for and optimization of electrode compositions based on abundant and cheap chemical elements. In this context, research in the field of Na-ion battery materials is acquiring increasing importance, mainly due to the fact that, in line with the envisaged increasing demand of rechargeable batteries to implement electric vehicles, as well as portable and stationary applications, concerns have arisen with regard to the cost and availability of lithium in the near future [1,2]. The ability to replace or, better, to complement Li-ion technology with different metal ion chemistry, in particular Na-ions, would lower costs and alleviate such concerns [3,4]. In addition, the use of Na instead of Li metal allows the use of aluminum as an anode current collector, providing a cost-effective alternative to copper [2].

Mn-based compounds are being widely investigated as cathode materials for rechargeable Na-ion batteries and, in particular, sodium manganese oxides such as $Na_{0.67}MnO_2$, $Na_{0.44}MnO_2$ and other Na compounds ($Na_xMn_yB_zO_2$, B = transition metal) have been regarded as promising cathode materials due to their high capacity, low cost and non-toxicity [5,6]. Among various cathode materials for Na-ion batteries, $Na_{0.44}MnO_2$ has been identified as highly promising because of its large capacity (theoretically 121 mAh g^{-1}) and good stability [7], even in aqueous electrolyte [8,9].

$Na_{0.44}MnO_2$ (NMO) is characterized by a peculiar tunnel-like structure composed of Mn-based octahedra and square-based pyramids connected through a vertex to form small, distorted hexagonal channels and larger S-shaped tunnels running parallel to the *c* crystallographic axis (Figure 1). Three partially occupied Na sites are present within the structure. The Na-cations of all sites can be shuttled during the electrochemical redox process, but to different extents [10,11]. The high reversibility of the insertion and de-insertion processes is only guaranteed in the compositional range $Na_{0.25}MnO_2$–$Na_{0.65}MnO_2$; otherwise, the tunnel structure is not preserved [12].

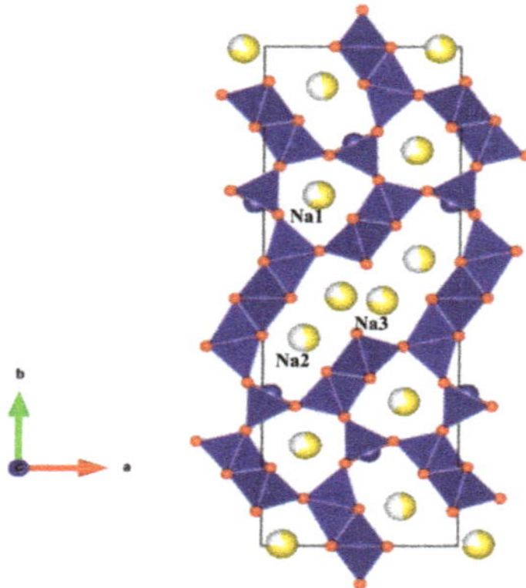

Figure 1. Crystal structure of $Na_{0.44}MnO_2$ showing the partial site occupancy on the Na sites as partially colored spheres.

NMO powders are conventionally prepared by solid-state reaction at 800 °C or higher for at least 9 h [10,13–15]. More recently, NMO materials have also been prepared by other synthesis procedures such as wet-chemistry techniques [12], including modified-Pechini methods [16–18], spray pyrolysis [19], polyvinilpyrrolidone (PVP)-assisted gel combustion synthesis [20,21], and reverse microemulsion methods [22,23]. However, even if a solution precursor is used, the NMO materials are usually obtained at high temperatures between 800 °C [12,17,18] and 950 °C [16,20,21] after several hours of annealing, typically ranging between 8 and 15 h. Notable exceptions are related to hydrothermal preparation strategies, where typically an aqueous solution is heated at only 205 °C, but for 4 days, in an autoclave [24,25]. Quite surprisingly, NMO materials deliver similar electrochemical performance and specific capacities at low current rates (i.e., approaching the theoretical capacity of 121 mA h g^{-1}), despite being synthesized via different methodologies.

In this work, we explore the possibility of reducing the temperature and duration of the thermal treatment necessary to synthesize the environmental friendly NMO from abundant raw materials through a solution combustion approach. This synthetic approach involves a self-sustained reaction in an aqueous solution containing metal nitrates as oxidizers and urea as fuel and reducing agent [26,27]. Indeed, we demonstrate that such an energy-saving synthesis yields a material with high electrochemical performance, reduced energy cost, and lower environmental impact.

2. Results and Discussion

2.1. Structural and Morphological Characterization of the as Prepared SC-Powder

Prior to the electrochemical characterization, all materials underwent a structural investigation to identify the appropriate synthesis conditions. Figure 2a shows the X ray diffraction patterns, XRDP, of the samples prepared with different Na/Mn ratios in the range 0.47–0.51, treated at 800 °C for 2 h. Results show that the sample prepared with a nominal ratio of 0.50 is single-phase, with the desired structure of $Na_{0.44}MnO_2$, i.e., an orthorhombic *P*bam space space group isostructural to $Na_4Mn_4Ti_5O_{18}$ (JCPDS: 01-076-0785) [28]. Samples with a lower Na/Mn ratio (0.47 and 0.49) contain Mn_2O_3 impurities (JCPDS: 01-71-0635), confirming a partial volatility of the alkali metal ions during the thermal treatment. Instead, the sample prepared with a higher Na content (Na/Mn ratio = 0.51) shows impurity peaks associated to the presence of α-$NaMnO_2$ (JCPDS: 01-25-0845).

Accordingly, the study on the required minimum temperature and time of the thermal treatment for obtaining single-phase $Na_{0.44}MnO_2$ was performed with a Na/Mn ratio of 0.50. Figure 2b reports the evolution of the diffraction patterns at different annealing temperatures. The XRDP show that at least 700 °C is required to achieve the preparation of phase-pure $Na_{0.44}MnO_2$ with the desired structure. Indeed, the mixture treated at 500 °C shows peaks related to the presence of crystalline Mn_2O_3 (JCPDS: 01-71-0635), weak peaks due to β-$Na_{0.70}MnO_2$ (JCPDS: 01-27-0752) and additional broader peaks of unidentified phases. After the thermal treatments at 600 °C, the dominant phase present is β-$Na_{0.70}MnO_2$, still partially detectable after the thermal treatment of 1 hour at 700 °C. Only after being treated at 700 °C for 2 h is the sample single-phase and able to be indexed according to the reflections of $Na_{0.44}MnO_2$, isostructural to $Na_4Mn_4Ti_5O_{18}$ (JCPDS: 01-76-0193).

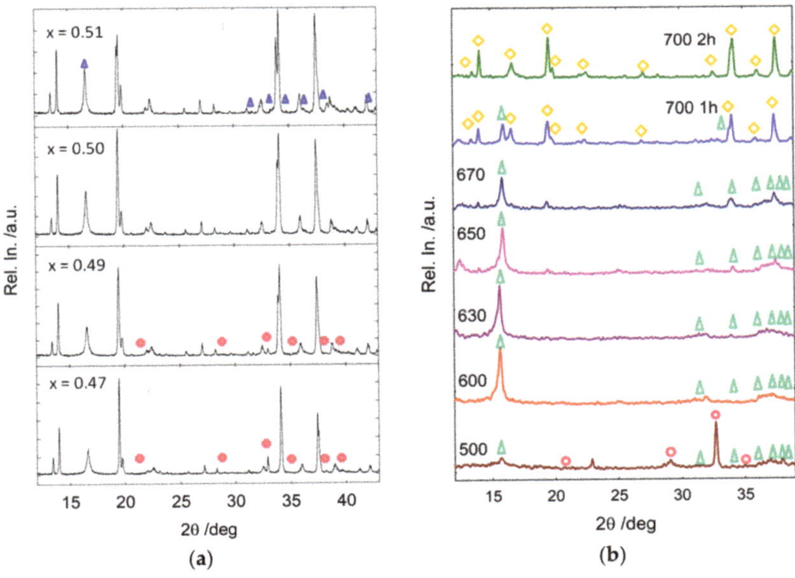

Figure 2. (**a**) X-Ray Diffraction Pattern, XRPD, of the samples prepared using different Na/Mn ratios in the precursor solution. Pink circles indicate the peaks of the Mn_2O_3 impurity while blue triangles those of α-$NaMnO_2$; (**b**) XRDP of the $Na_{0.44}MnO_2$ powders treated at increasing temperature. Pink circles indicate the peaks of the Mn_2O_3 impurity, green triangles those of β-$Na_{0.7}MnO_2$ and yellow squares are related to the $Na_{0.44}MnO_2$ phase.

Figure 3 presents the SEM images of the sample prepared according to optimized solution combustion synthesis conditions (SC-NMO) and, as comparison, of the sample prepared according the solid state route (SS-NMO). The scanning electron microscopy (SEM) images show that SC-NMO (Figure 3a,b) is mainly composed of sub-micrometric slabs of 100–400 nm width and 0.6–1.4 μm length. Hence, the particle morphology of SC-NMO is similar to that of $Na_{0.44}MnO_2$ reported in our previous work [29]. However, the dimensions of the SC-NMO slabs are smaller and surfaces of the slabs are less defined as compared to SS-NMO (Figure 3c). Heat treated $Na_{0.44}MnO_2$ samples generally present rod-like structures due to the preferential growth along the [0 0 1] direction [22]. The present finding is therefore in agreement with the reduced time and temperature of the thermal treatment for SC-NMO compared to SS-NMO. Indeed, as expected, the synthetic approach here proposed is a solution-based approach. Wet-chemistry techniques are well known to produce even rather complex oxides at relative low temperatures, compared to conventional solid-state synthesis, thanks to the homogeneous and intimate mixing of the metal ions in the liquid phase and the formation of a porous matrix composed of nanoparticles that can easily react at the solid state in subsequent thermal treatments [30]. In addition, compared to other wet-chemistry techniques, the solution combustion approach takes advantage of the exothermicity of the redox reaction involved in the precursor formation; the released heat of the combustion reaction may fulfill, at least locally, the energy requirement for the formation of the oxide matrix, thus allowing the preparation of single-phase materials at relatively lower temperatures and in a shorter time [26].

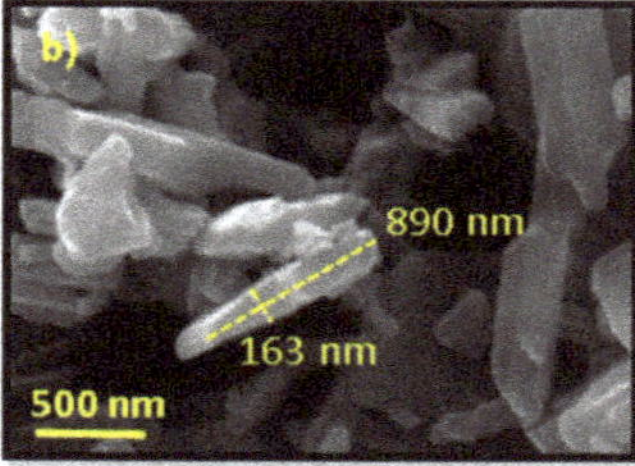

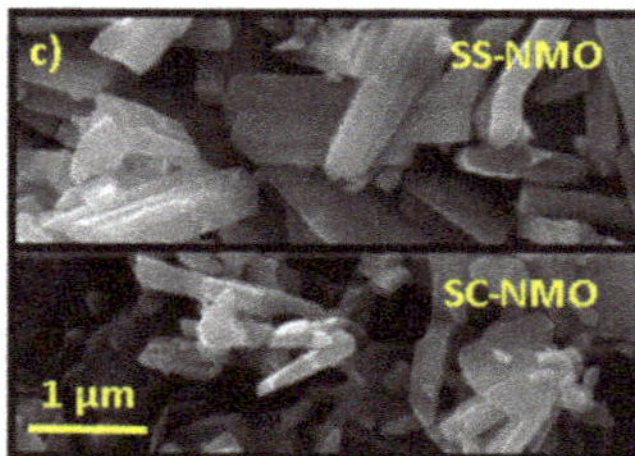

Figure 3. Scanning Electron Microscopy, SEM, micrographs of the SC-NMO as synthesized powder. The magnifications are: (**a**) 25.0 kx and (**b**) 100.0 kx; (**c**) Comparison of two micrographs acquired at the same SEM mag. (50.0 kx): the upper image refers to SS-NMO sample; the lower one to SC-NMO sample.

2.2. Electrochemical Tests

Figure 4 shows the cyclic voltammetry (CV) scans of the SC-NMO electrode between 2.0 V–3.8 V. The general shape of the voltammograms, including the position of the current peaks, is in very good agreement with the literature. In detail, seven main anodic peaks (at 2.23, 2.49, 2.70, 3.00, 3.10, 3.23 and 3.45 V) and cathodic peaks (3.43, 3.22, 3.07, 2.97, 2.66, 2.43 and 2.19 V) are evident, confirming the reversibility of the process. Furthermore, each current peak is correlated to the (de)insertion of sodium from a particular crystallographic site [10,11]. Differently from what was observed with other morphologies, such as in NMO nanofibers [31], SC-NMO does not need an activation cycle: in fact, the current peaks of the first cyclic sweep are overlapping with those of the following cycles.

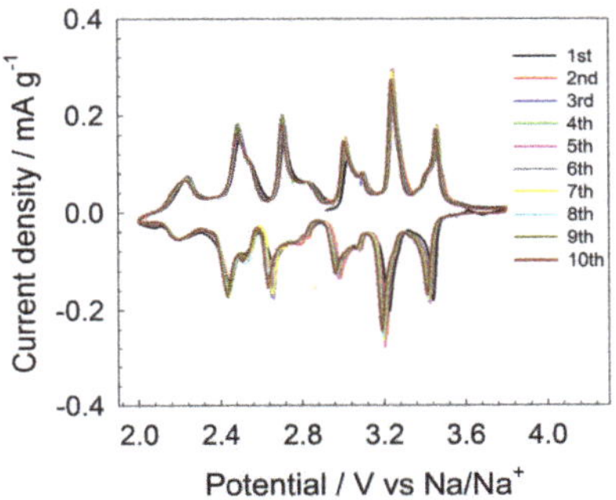

Figure 4. Cyclic Voltammetry, CV, of the SC-NMO tested for 10 cycles in the potential range 2.0 V–3.8 V with the electrolyte $NaPF_6$ 1M in PC.

Figure 5 reports the cycling performance of the SC-NMO at increasing C rates (C/10, C/5, C/2, 1 C, 2 C and 5 C) and during the long-term cycling test at 1 C. In addition, selected potential profiles of SC-NMO, recorded at different C-rates and during the 1st and 2nd cycle at C/10 are also given. The galvanostatic cycling data are compared with SS-NMO to reveal the impact of the modified synthesis conditions [29]. The capacities at C/10 are approximately 115 mA h g^{-1}, i.e., in very good agreement with literature data and close to the theoretical capacity of 121 mA h g^{-1}. For example, at the same C rate, a reversible capacity of 110 mA h g^{-1} was reported for a sample prepared through solid state reaction at 800 °C for 9 h [14] and 112 mA h g^{-1} for a sample prepared through a solution synthesis approach followed by a thermal treatment at 800 °C for 10 h [32].

At low current densities, such as C/10 or C/5, the SS-NMO-based electrode delivers only a slightly higher capacity of about 1–2 mAh g^{-1}, both in charge and in discharge, which is within the margin of experimental error of the battery cycles used. However, at C/2, both materials show the same performance, and from 1 C on the SC-NMO outperforms SS-NMO in terms of delivered specific capacity and coulombic efficiency. At 1 C the difference is still small but at 2 C the capacity difference between the two materials increases to 10 mA h g^{-1}, i.e., SC-NMO delivering 95 mA h g^{-1} and SS-NMO only 85 mA h g^{-1}. Both materials show a rather reversible behavior after the current rate test, i.e., when reducing the cycling rate at C/10.

It should be noted that the improved performance of SC-NMO at high C rates is not related to residual carbon coating, because the preparation of SC-NMO is performed in oxidizing atmosphere, i.e., air, to avoid the material's degradation. In fact, even in a slightly reducing atmosphere, such as nitrogen, the material degrades to a mixture of Mn_2O_3 and Na_2O, which readily transforms into $NaCO_3$ under ambient conditions (see Figure S1, Electronic Supplementary Information (ESI)). Consistently, the TGA traces (Figure S2, ESI) do not indicate for the presence of additional carbon. In fact, the degree of weight loss for the SS-NMO sample, prepared in the absence of organic reagents, is even higher than that pertaining to the SC-NMO sample during a first acquisition of the TGA curve.

The potential profiles during discharge and charge for the 3rd cycle at each C rate are reported in Figure 5b,c, respectively. The plateaus in the potential profiles correspond well with the current peaks observed via cyclic voltammetry in Figure 4. SC-NMO and SS-NMO were also cycled at 1 C for 200 cycles (Figure 5e) and selected potential profiles of SC-NMO upon cycling are shown in Figure 5f. After 100 cycles, 2 cycles at C/10 were performed to evaluate material degradation. As can be clearly seen in Figure 5e, the capacity of the SC-NMO material is always higher than the one of the SS. After 200 cycles the discharge capacity of SC-NMO is 73.8 mA h g^{-1} with a coulombic efficiency of 99.72% and a capacity retention of 75.7%. Interestingly, the SC-NMO capacity is completely retained (around 108 mA h g^{-1}) when current is changed to C/10 after the long-term cycling at 1 C. This suggests that the fading performance during long-term cycling is correlated with the composite electrode degradation rather than the irreversible structural degradation of the active material.

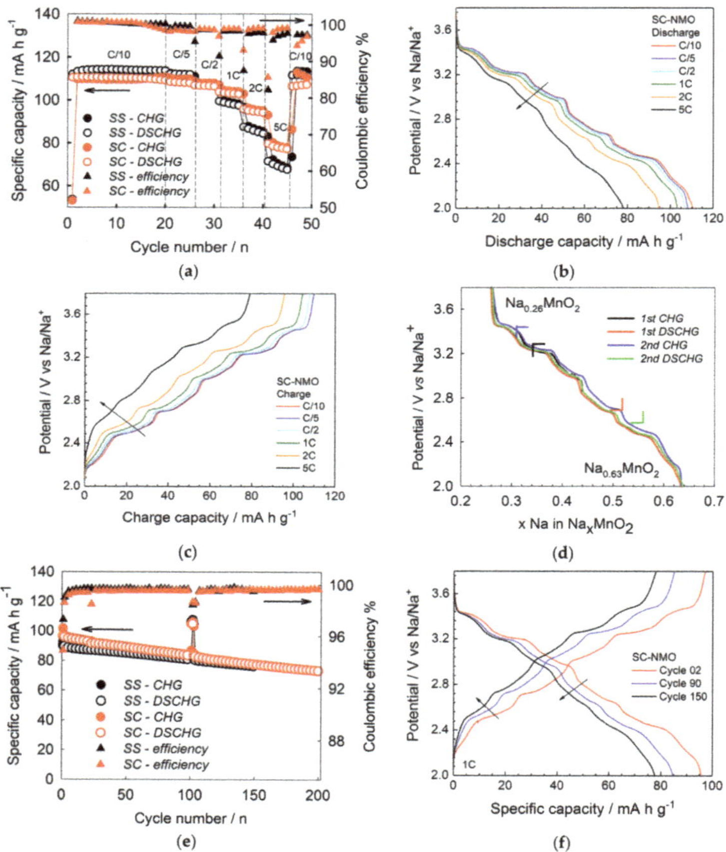

Figure 5. (**a**) Comparison of the C rate tests performed on the SS-NMO-based electrode [29] and the SC-NMO-based electrode presented in this work: the tested C rates are: C/10, C/5, C/2, 1 C, 2 C, 5 C; (**b**,**c**) show the potential profiles of SC-NMO, respectively, during the 3rd discharge and charge cycle at each C rate; (**d**) "Potential vs. x Na in Na_xMnO_2" graph reported for the 1st and 2nd cycles at C/10. The arrows show if the material is increasing or decreasing its sodium content while cycling; (**e**) 1 C long-term cycling comparison between SC-NMO-based electrode (200 cycles) and SS-NMO-based electrode (150 cycles—reported in our previous work [29]); (**f**) selected potential profiles of SC-NMO upon cycling (panel **e**).

The high coulombic efficiency registered for the SC-NMO again suggests a highly reversible insertion of Na^+ ions. Indeed, as reported in Figure 5d, the amount of Na^+ equivalents in SC-NMO is between 0.63 eq. Na^+ and 0.26 eq. Na^+, i.e., well within the range where high reversibility is observed [13]. As a matter of fact, the potential profiles of the first and second charge plotted vs. Na^+ equivalents (Figure 5d) almost overlap.

3. Materials and Methods

3.1. Solution Combustion Synthesis of $Na_{0.44}MnO_2$

The powder sample of nominal composition $Na_{0.44}MnO_2$ (solution combustion synthesis $Na_{0.44}MnO_2$, SC-NMO) was prepared starting from appropriate amounts of $NaNO_3$, $Mn(NO_3)_2.4H_2O$ and urea, $CO(NH_2)_2$ (Sigma-Aldrich S.r.l, Milano, Italy). The chemicals were dissolved in deionized water in the appropriate stoichiometric ratio and heated at 80 °C under continuous stirring for 30 min. The temperature was then increased to 100 °C to achieve complete evaporation of the solvent until a viscous light pink resin was obtained, which spontaneously ignited to produce a black precursor. This latter was subsequently treated in a furnace at a temperature between 500 °C and 700 °C for maximum 2 h. The Na/Mn ratio was modified between 0.47–0.51 to identify the Na excess suitable to compensate for the loss of alkali metal during the high temperature synthesis steps.

For comparison, a sample of the same nominal composition (solid-state synthesis $Na_{0.44}MnO_2$, SS-NMO) was prepared according to the solid-state synthesis procedure reported by Sauvage et al. [13], which foresees a 10 wt.% excess of Na precursor, corresponding to a starting stoichiometry of $Na_{0.49}MnO_2$, to compensate for the Na ion volatility at high temperature. The characterization of this sample was already described in our previous work [29].

3.2. Morphological and Structural Characterization of SC-NMO

SC-NMO powder was sputter-coated with Au and analyzed through Scanning Electron Microscopy (SEM, Mira 3, Tescan, Brno, Czech Republic), using a beam acceleration voltage of 20 kV. The average dimensions of the slabs were calculated by measuring the width and length of at least 75 single slabs through the ImageJ® free software. X-ray Diffraction Patterns (XRDP, BRUKER diffractometer D8 Advance, Cu-Kα: λ = 0.154 nm, Bruker Italia S.r.l., Milano, Italy) were acquired on the SC-NMO powder using an Al sample holder. The data were acquired in the 2θ range from 10° to 45° with scan step of 0.02 and a fixed counting time per step of 8 s. The powder underwent a Thermo-Gravimetric Analysis (TGA, Q5000, TA Instruments, New Castle, DE, USA) under static air, with a heating ramp of 10 °C/min from room temperature up to 800 °C.

3.3. Casting of the Electrodes

The NMO (both SC or SS) powders were mechanically sieved (mesh size $\leq$ 45 μm), after hand-grinding to separate possible agglomerates of particles. The active material, the conductive carbon (SuperC65®, IMERYS, Switzerland) and binder polyvinylidene fluoride (PVdF, Solef 6020, Solvay Polymer specialties, Bruxelles, Belgium) were mixed via stirring overnight in N-methyl-2-pyrrolidone (NMP, Sigma Aldrich) in the weight ratio 85:10:5, respectively. In detail, 1 mL of NMP was used for 0.588 g of solid (solid content of 34.8%). Slurries with a wet thickness of 150 μm were cast on aluminum foil (Hydro, 0.020 mm thick). The slurries were dried overnight in an oven at 60 °C, and then round disc electrodes with a diameter of 12 mm-diameter were cut. The electrodes were dried under vacuum in a glass oven (Büchi) for 3 h at room temperature, 3 h at 60 °C and finally for 12 h at 100 °C. Subsequently, electrodes were pressed (8 ton for 10 s), weighed and dried again under vacuum for 3 h at room temperature and 8 h at 100 °C. The final electrodes had an average active material mass loading of 2.7 to 3.2 mg cm^{-2}.

3.4. Cell Assembly and Electrochemical Tests

Three-electrode Swagelok cells were assembled inside an argon-filled glove-box (mBraun) with water and oxygen content below 0.1 ppm. The cell configuration for all tests was: Na metal (99.8%, ACROS ORGANICS, Geel, Belgium) as reference and counter electrode (RE, CE), 1M $NaPF_6$ (sodium hexafluorophosphate, 99+%, FluoroChem, Glossop, UK) in PC (propylene carbonate, battery grade, BASF, Ludwigshafen, Germany) as electrolyte with glass fiber disks (GF/D, Whatman, Maidstone, UK) as separator and NMO as working electrode (WE). $NaPF_6$ and PC were used without further purification.

Electrodes (all possessing similar mass loading) were tested through cyclic voltammetry (CV) with a scan rate of 0.1 mV s^{-1} in the potential range of 2.0 V–3.8 V vs. Na/Na^+ (VMP3 multichannel potentiostat, BioLogic, Seyssinet-Pariset, France). The galvanostatic cycling was performed between 2.0 V and 3.8 V at various current densities ranging from C/10 to 5 C (1 C = 121 mA g^{-1}) (Maccor battery tester 4300, Maccor Inc., OK, USA). All electrochemical tests were carried out at 20 ± 1 °C.

4. Conclusions

For the first time, $Na_{0.44}MnO_2$ nanometric slabs were synthesized using an urea-based solution combustion synthesis and a subsequent annealing process. This simple synthetic approach, which makes use of the eco-friendly urea, enabled the preparation of $Na_{0.44}MnO_2$ cathode material in less time (2 h) and at a lower annealing temperature (700 °C) compared to state-of-the-art synthesis routes such as solid-state and sol-gel methods. SEM revealed smaller particle dimensions as a result of the aforementioned synthesis conditions. Moreover, the final cathode material has shown superior electrochemical performances, in terms of delivered capacities at high C rates of up to 5 C (i.e., 105 mAh g^{-1} at 1 C and 80 mAh g^{-1} at 5 C).

The improved capacities at high rates can be ascribed to the different morphology, i.e., smaller slabs characterized by a lower degree of anisotropy. Hence, the results suggest that reducing the particle sizes may be a suitable strategy for improving the high rate performances of $Na_{0.44}MnO_2$ cathode materials.

Supplementary Materials: The following are available online at www.mdpi.com/2313-0105/4/1/8/s1, Figure S1: XRDP of the NMO sample after a thermal treatment at 800 °C under N2 flux, Figure S2: Thermogravimetric analysis of the SS- and SC-NMO powders.

Acknowledgments: L.G.C., S.P and D.B acknowledge the financial support of the Helmholtz Association. C.T. is grateful for the financial support of Fondazione Cariplo and Regione Lombardia through grant 2015-0753.

Conflicts of Interest: The authors declare no conflict of interest.

References

1. Kundu, D.; Talaie, E.; Duffort, V.; Nazar, L.F. The Emerging Chemistry of Sodium Ion Batteries for Electrochemical Energy Storage. *Angew. Chem. Int. Ed.* **2015**, *54*, 3432–3448. [CrossRef] [PubMed]
2. Hwang, J.-Y.; Myung, S.-T.; Sun, Y.-K. Sodium-ion batteries: Present and future. *Chem. Soc. Rev.* **2017**, *46*, 3529–3614. [PubMed]
3. Kubota, K.; Komaba, S. Review—Practical Issues and Future Perspective for Na-Ion Batteries. *J. Electrochem. Soc.* **2015**, *162*, A2538–A2550. [CrossRef]
4. Yabuuchi, N.; Kubota, K.; Dahbi, M.; Komaba, S. Research Development on Sodium-Ion Batteries. *Chem. Rev.* **2014**, *114*, 11636–11682. [CrossRef] [PubMed]
5. Kim, H.; Kim, H.; Ding, Z.; Lee, M.H.; Lim, K.; Yoon, G.; Kang, K. Recent Progress in Electrode Materials for Sodium-Ion Batteries. *Adv. Energy Mater.* **2016**, *6*, 1600943.
6. Xiang, X.; Zhang, K.; Chen, J. Recent Advances and Prospects of Cathode Materials for Sodium-Ion Batteries. *Adv. Mater.* **2015**, *27*, 5343–5364. [CrossRef] [PubMed]

7. Cao, Y.; Xiao, L.; Wang, W.; Choi, D.; Nie, Z.; Yu, J.; Saraf, L.V.; Yang, Z.; Liu, J. Reversible Sodium Ion Insertion in Single Crystalline Manganese Oxide Nanowires with Long Cycle Life. *Adv. Mater.* **2011**, *23*, 3155–3160. [CrossRef] [PubMed]
8. Hung, T.F.; Lan, W.H.; Yeh, Y.W.; Chang, W.S.; Yang, C.C.; Lin, J.C. Hydrothermal Synthesis of Sodium Titanium Phosphate Nanoparticles as Efficient Anode Materials for Aqueous Sodium-Ion Batteries. *ACS Sustain. Chem. Eng.* **2016**, *4*, 7074–7079. [CrossRef]
9. Li, Z.; Young, D.; Xiang, K.; Carter, W.C.; Chiang, Y.M. Towards High Power High Energy Aqueous Sodium-Ion Batteries: The $NaTi_2(PO_4)_3/Na_{0.44}MnO_2$ System. *Adv. Energy Mater.* **2013**, *3*, 290–294. [CrossRef]
10. Ma, G.; Zhao, Y.; Huang, K.; Ju, Z.; Liu, C.; Hou, Y.; Xing, Z. Effects of the Starting Materials of $Na_{0.44}MnO_2$ Cathode Materials on Their Electrochemical Properties for Na-Ion Batteries. *Electrochim. Acta* **2016**, *222*, 36–43. [CrossRef]
11. Kim, H.; Kim, D.J.; Seo, D.H.; Yeom, M.S.; Kang, K.; Kim, D.K.; Jung, Y. Ab Initio Study of the Sodium Intercalation and Intermediate Phases in $Na_{0.44}MnO_2$ for Sodium-Ion Battery. *Chem. Mater.* **2012**, *24*, 1205–1211. [CrossRef]
12. Zhou, X.; Guduru, R.K.; Mohanty, P. Synthesis and Characterization of $Na_{0.44}MnO_2$ from Solution Precursors. *J. Mater. Chem. A* **2013**, *1*, 2757–2761. [CrossRef]
13. Sauvage, F.; Laffont, L.; Tarascon, J.M.; Baudrin, E. Study of the Insertion/deinsertion Mechanism of Sodium into $Na_{0.44}MnO_2$. *Inorg. Chem.* **2007**, *46*, 3289–3294. [CrossRef] [PubMed]
14. Zhao, L.; Ni, J.; Wang, H.; Gao, L. $Na_{0.44}MnO_2$–CNT Electrodes for Non-Aqueous Sodium Batteries. *RSC Adv.* **2013**, *3*, 6650–6655. [CrossRef]
15. Wang, C.H.; Yeh, Y.W.; Wongittharom, N.; Wang, Y.C.; Tseng, C.J.; Lee, S.W.; Chang, W.S.; Chang, J.K. Rechargeable $Na/Na_{0.44}MnO_2$ Cells with Ionic Liquid Electrolytes Containing Various Sodium Solutes. *J. Power Sources* **2015**, *274*, 1016–1023. [CrossRef]
16. Xu, M.; Niu, Y.; Chen, C.; Song, J.; Bao, S.; Li, C.M. Synthesis and Application of Ultra-Long $Na_{0.44}MnO_2$ Submicron Slabs as a Cathode Material for Na-Ion Batteries. *RSC Adv.* **2014**, *4*, 38140–38143. [CrossRef]
17. Kim, D.J.; Ponraj, R.; Kannan, A.G.; Lee, H.W.; Fathi, R.; Ruffo, R.; Mari, C.M.; Kim, D.K. Diffusion Behavior of Sodium Ions in $Na_{0.44}MnO_2$ in Aqueous and Non-Aqueous Electrolytes. *J. Power Sources* **2013**, *244*, 758–763. [CrossRef]
18. Ruffo, R.; Fathi, R.; Kim, D.J.; Jung, Y.H.; Mari, C.M.; Kim, D.K. Impedance Analysis of $Na_{0.44}MnO_2$ Positive Electrode for Reversible Sodium Batteries in Organic Electrolyte. *Electrochim. Acta* **2013**, *108*, 575–582. [CrossRef]
19. Shen, K.Y.; Miklos, L.; Wang, L.; Axelbaum, R.L. Spray Pyrolysis and Electrochemical Performance of $Na_{0.44}MnO_2$ for Sodium-Ion Battery Cathodes. *MRS Commun.* **2017**, *7*, 74–77. [CrossRef]
20. Qiao, R.; Dai, K.; Mao, J.; Weng, T.C.; Sokaras, D.; Nordlund, D.; Song, X.; Battaglia, V.S.; Hussain, Z.; Liu, G.; et al. Revealing and Suppressing Surface Mn(II) Formation of $Na_{0.44}MnO_2$ Electrodes for Na-Ion Batteries. *Nano Energy* **2015**, *16*, 186–195. [CrossRef]
21. Dai, K.; Mao, J.; Song, X.; Battaglia, V.; Liu, G. $Na_{0.44}MnO_2$ with Very Fast Sodium Diffusion and Stable Cycling Synthesized via Polyvinylpyrrolidone-Combustion Method. *J. Power Sources* **2015**, *285*, 161–168. [CrossRef]
22. Zhan, P.; Wang, S.; Yuan, Y.; Jiao, K.; Jiao, S. Facile Synthesis of Nanorod-like Single Crystalline $Na_{0.44}MnO_2$ for High Performance Sodium-Ion Batteries. *J. Electrochem. Soc.* **2015**, *162*, A1028–A1032. [CrossRef]
23. Liu, Q.; Hu, Z.; Chen, M.; Gu, Q.; Dou, Y.; Sun, Z.; Chou, S.; Dou, S.X. Multiangular Rod-Shaped $Na_{0.44}MnO_2$ as Cathode Materials with High Rate and Long Life for Sodium-Ion Batteries. *ACS Appl. Mater. Interfaces* **2017**, *9*, 3644–3652. [CrossRef] [PubMed]
24. Hosono, E.; Matsuda, H.; Honma, I.; Fujihara, S.; Ichihara, M.; Zhou, H. Synthesis of Single Crystalline Electro-Conductive $Na_{0.44}MnO_2$ Nanowires with High Aspect Ratio for the Fast Charge-Discharge Li Ion Battery. *J. Power Sources* **2008**, *182*, 349–352. [CrossRef]
25. Hosono, E.; Saito, T.; Hoshino, J.; Okubo, M.; Saito, Y.; Nishio-Hamane, D.; Kudo, T.; Zhou, H. High Power Na-Ion Rechargeable Battery with Single-Crystalline $Na_{0.44}MnO_2$ Nanowire Electrode. *J. Power Sources* **2012**, *217*, 43–46. [CrossRef]
26. Li, F.; Ran, J.; Jaroniec, M.; Qiao, S.Z. Solution combustion synthesis of metal oxide nanomaterials for energy storage and conversion. *Nanoscale* **2015**, *7*, 17590–17610. [CrossRef] [PubMed]

27. Aruna, S.T.; Mukasyan, A.S. Combustion synthesis and nanomaterials. *Curr. Opin. Solid State Mater. Sci.* **2008**, *12*, 44–50. [CrossRef]
28. Mumme, W. The Structure of $Na_4Mn_4Ti_5O_{18}$. *Acta Crystallogr. Sect. B Struct. Sci. Cryst. Eng. Mater.* **1968**, *24*, 1114–1120. [CrossRef]
29. Dall'Asta, V.; Buchholz, D.; Chagas, L.G.; Dou, X.; Ferrara, C.; Quartarone, E.; Tealdi, C.; Passerini, S. Aqueous Processing of $Na_{0.44}MnO_2$ Cathode Material for the Development of Greener Na-Ion Batteries. *ACS Appl. Mater. Interfaces* **2017**, *9*, 34891–34899. [CrossRef] [PubMed]
30. Qiao, L.; Swihart, M.T. Solution-phase synthesis of transition metal oxide nanocrystals: Morphologies, formulae, and mechanisms. *Adv. Coll. Interface Sci.* **2017**, *244*, 199–266. [CrossRef] [PubMed]
31. Fu, B.; Zhou, X.; Wang, Y. High-Rate Performance Electrospun $Na_{0.44}MnO_2$ Nanofibers as Cathode Material for Sodium-Ion Batteries. *J. Power Sources* **2016**, *310*, 102–108. [CrossRef]
32. He, X.; Wang, J.; Qiu, B.; Paillard, E.; Ma, C.; Cao, X.; Liu, H.; Stan, M.C.; Liu, H.; Gallash, T.; et al. Durable High-Rate Capability $Na_{0.44}MnO_2$ Cathode Material for Sodium-Ion Batteries. *Nano Energy* **2016**, *27*, 602–610. [CrossRef]

Review

A Review of Model-Based Design Tools for Metal-Air Batteries

Simon Clark [1,2], Arnulf Latz [1,2,3] and Birger Horstmann [1,2,*]

[1] German Aerospace Center (DLR), Pfaffenwaldring 38-40, 70569 Stuttgart, Germany; simon.clark@dlr.de (S.C.); arnulf.latz@dlr.de (A.L.)
[2] Helmholtz Institute, Ulm University (UUlm), Helmholtzstr 11, 89081 Ulm, Germany
[3] Institute for Electrochemistry, Ulm University (UUlm), Albert-Einstein-Allee 47, 89081 Ulm, Germany
[*] Correspondence: birger.horstmann@dlr.de; Tel.: +49-(0)731-50-34311

Received: 8 December 2017; Accepted: 19 January 2018; Published: 29 January 2018

Abstract: The advent of large-scale renewable energy generation and electric mobility is driving a growing need for new electrochemical energy storage systems. Metal-air batteries, particularly zinc-air, are a promising technology that could help address this need. While experimental research is essential, it can also be expensive and time consuming. The utilization of well-developed theory-based models can improve researchers' understanding of complex electrochemical systems, guide development, and more efficiently utilize experimental resources. In this paper, we review the current state of metal-air batteries and the modeling methods that can be implemented to advance their development. Microscopic and macroscopic modeling methods are discussed with a focus on continuum modeling derived from non-equilibrium thermodynamics. An applied example of zinc-air battery engineering is presented.

Keywords: metal-air; zinc-air; modeling; simulation; computational chemistry

---

1. Introduction

In the ever-growing search for safe and high-performance energy storage technology, development of metal-air batteries is taking on new importance [1]. The promise of these systems is clear: a significant increase in energy density over Li-ion batteries, utilization of abundant materials, and improved safety [2]. While great progress has been made in their development, challenges remain before secondary metal-air batteries can become widely commercially viable.

Metal-air batteries comprise a metal electrode (e.g., Zn, Li, Mg, Al, etc.), electrolyte (aqueous or non-aqueous), and a bi-functional air electrode (BAE). The basic operating principle is to electrochemically reduce O_2 from air and oxidize the metal electrode to reversibly form solid metal-oxides. In this way, both the volume and the weight of the battery can be significantly reduced compared to Li-ion systems. Figure 1 compares the theoretical energy density and specific energy of metal-air systems. In some non-ideal cases the precipitation of the solid discharge product can consume active electrolyte components, reducing the achievable energy density [3]. Research into a variety of metal-air chemistries is ongoing. The homogeneous deposition of Mg metal makes Mg-air systems appealing [4–6], but aqueous Mg-air batteries are severely limited by the corrosion of the Mg electrode. Ionic liquid electrolytes have been proposed for Mg-air systems, but they also suffer from electrochemical instability, particularly during charging, and the reversibility of the cell is limited [7]. Another interesting contender is Al-air. Al is an abundant and safe material, and Al-air batteries have high theoretical energy density and specific energy values [8–10]. However theses systems are susceptible to corrosion and have not demonstrated adequate cycling stability. The natural abundance and safety of sodium combined with its comparable properties with lithium have driven research into Na-air [11–13]. These systems are still in a very early stage of research. Si-air batteries have also attracted attention [14]. They have a high theoretical energy density and are stable in aqueous electrolytes. Experimental studies of Si-air systems have been performed in both ionic liquid [15] and alkaline electrolytes [16], but they currently face challenges with the reversibility of the solid discharge product, precipitation, and pore blockage. Among the metal-air systems under development, Li-air and Zn-air are the most promising [17–19].

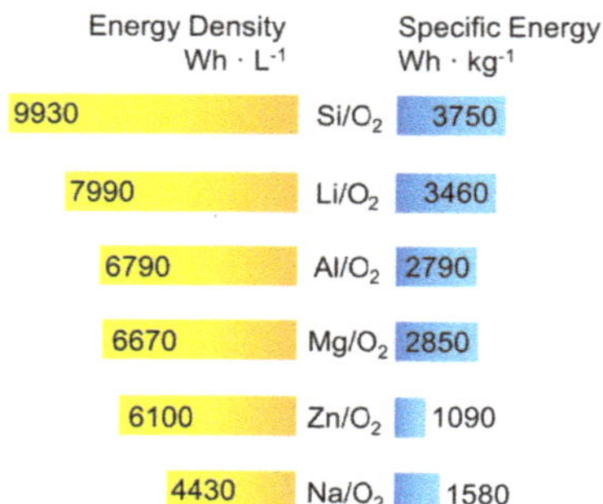

Figure 1. Overview of the theoretical energy density and specific energy (including oxygen) of commonly researched metal-air batteries. Values are calculated considering the specific mass and volume of the discharge product with the OCV and charge transfered in the cell reaction.

Li-air batteries (LABs) have been researched for decades [20], but have only become a widely-pursued topic since the early 2000s. The electrolyte has proved to be a limiting factor in LAB development. The most common electrolyte configurations of Li-air systems are aprotic (non-aqueous) and aqueous. Mixed electrolyte systems have also been proposed. The beginning of the LAB research wave focused on aprotic electrolytes. The first work on the aprotic Li-air system ($LiPF_6$ in ethylene carbonate (EC)) was performed in 1996 by Abraham, et al. [21], who proposed an overall reaction forming Li_2O_2 or Li_2O. Early aprotic Li-air cells were based on a carbonate solvent, but it has since been shown that carbonate solvents are unstable, producing lithium-carbonates during discharging and evolving CO_2 during charing [22–25]. These days, carbonate electrolytes have been abandoned in favor of ether and ester solvents with lithium salts. A second challenge for LABs in aprotic electrolytes relates to the precipitation of Li_2O_2. When this solid precipitates in the cathode, it can form a dense layer over the carbon surface and inhibit the transfer of electrons. As precipitation continues, entire pores in the cathode may become blocked, passivating the electrode and limiting the cell capacity. Finally, it has been noted that oxygen transport in aprotic eletrolytes can be a challenging factor in LAB performance, especially at higher current densities [26,27]. This has motivated researches to learn from the success of the gas diffusion electrode in fuel cells and pursue investigations of LABs with aqueous electrolytes.

It is well known that Li metal reacts violently with water, which had previously limited the use of aqueous electrolytes for Li-air systems. Then in 2004, a glass ceramic layer over the Li electrode was successfully proposed to protect the metal electrode while still allowing the electrochemical reaction to proceed [28,29]. In alkaline aqueous electrolytes, the discharge product is $LiOH \cdot H_2O$ instead of Li_2O_2. In these systems, $LiOH \cdot H_2O$ tends to precipitate at the separator-anode interface [30], which reduces the risk of pore clogging in the cathode as observed in aprotic LABs. However, when aqueous alkaline electrolytes are exposed to air, dissolved CO_2 reacts with OH^- to form carbonates, which slowly reduces the conductivity of the electrolyte and limits the lifetime of the cell.

In the recent wave of interest in the development of Li-air batteries, many theoretical studies have highlighted the impressive possibilities of these systems [30–36] and the company International

Business Machines Corporation (IBM) pursued Li-air systems for commercial applications [17]. Although significant challenges remain [37–41], the future of Li-air batteries is promising.

Zn-air batteries (ZABs) stand alone as the only fully mature metal-air system and have been successfully commercialized as primary cells for many years. They are particularly suitable for low-current applications like hearing aids. However, their calendar life and electrical rechargeability is limited [42]. One major advantage of Zn as an electrode material is that, unlike Li, it is stable in water. In an effort to improve the rechargeability of ZABs, alternative near-neutral aqueous [43,44] and ionic liquid electrolytes [45–47] have been proposed. Current research continues to focus on material development to address the lifetime limitations and cell engineering to improve the performance of these systems [48]. Although some hurdles remain, the development of secondary ZABs has progressed to the point that they could become feasible for stationary storage applications and some Start-Ups like Eos Energy Storage [49] and Fluidic Energy [50] have begun to commercialize the technology. Zn-air systems offer perhaps the most immediate and reliable pathway to a viable secondary metal-air battery.

In this review, we examine model-based design tools that can be applied to advance development of metal-air systems. The majority of existing models were developed for Zn-air cells, as they are the oldest and most mature system. Therefore our review shall focus mainly on the application of models to ZABs and highlight some important advances unique to LABs. Note that these methods are easily translatable to other metal-air systems.

2. Zinc-Air Batteries

In this section, we present a summary of the working principle of ZABs and discuss the main challenges hindering the development of electrically-rechargeable Zn-air systems.

2.1. Working Principle

In their most common configuration, Zn-air batteries contain a metallic Zn electrode, porous separator, circa 30 wt % aqueous KOH electrolyte, and a bi-functional air electrode (BAE). The BAE consists of a porous substrate and a bi-functional air catalyst (e.g., MnO_2) to facilitate the oxygen-reduction and oxygen-evolution reactions (ORR, OER) [51–53]. The design of the BAE is similar to gas diffusion layers (GDL) from fuel cell applications. The porous BAE substrate contains carbon fibers and binder with mixed hydrophilic and hydrophobic properties to promote the formation of the so-called three phase boundary, while hindering the electrolyte from flooding out. The Zn electrode is often a paste consisting of zinc metal powder, electrolyte, and binder [54]. The capacity of the cell is determined by the Zn electrode, which is designed in such a way as to include as much active material as possible while minimizing the effects of shape-change and electrolyte concentration gradients. Aqueous KOH is the most common electrolyte for Zn-air batteries due to its high conductivity (circa 600 $\mathrm{mS \cdot cm^{-1}}$).

An operational schematic of a ZAB in alkaline electrolyte is shown in Figure 2. When the cell is discharged, the Zn electrode is electrochemically oxidized to form $Zn(OH)_4{}^{2-}$ (zincate) complexes,

$$\mathrm{Zn + 4OH^- \rightleftharpoons Zn(OH)_4{}^{2-} + 2\,e^-}, \qquad \mathrm{E^0 = -1.199\ V\ v.\ RHE}. \tag{1}$$

Oxygen gas enters the cell through the BAE and dissolves into the electrolyte, where it is reduced to form OH^- ions,

$$\mathrm{0.5\,O_2(aq) + H_2O + 2\,e^- \rightleftharpoons 2\,OH^-}, \qquad \mathrm{E^0 = 0.401\ V\ v.\ RHE}. \tag{2}$$

When the saturation limit of zinc in the electrolyte is exceeded, solid ZnO precipitates mainly in the anode and the battery achieves a stable working point,

$$\mathrm{Zn(OH)_4{}^{2-} \rightleftharpoons ZnO(s) + H_2O + 2OH^-}. \tag{3}$$

When the cell is charged, ZnO dissolves, zinc is redeposited at the Zn electrode and oxygen gas is evolved in the BAE. The overall cell reaction is given by

$$\mathrm{Zn} + 0.5\,\mathrm{O_2(aq)} \rightleftharpoons \mathrm{ZnO(s)}. \tag{4}$$

The open-circuit voltage of a ZAB in 30 wt % KOH is 1.65 V. The high conductivity of the electrolyte, high mobility of OH^-, and reasonable kinetics of the ORR give the cell a nominal discharge voltage of 1.2 V at current densities of circa 10 $mA \cdot cm^{-2}$ [55]. The molar volume of ZnO is 60% larger than metallic Zn, which causes the cell to expand during discharge. Even so, ZAB button cells demonstrate a practical energy density on the order of 1000 $Wh \cdot L^{-1}$ [55].

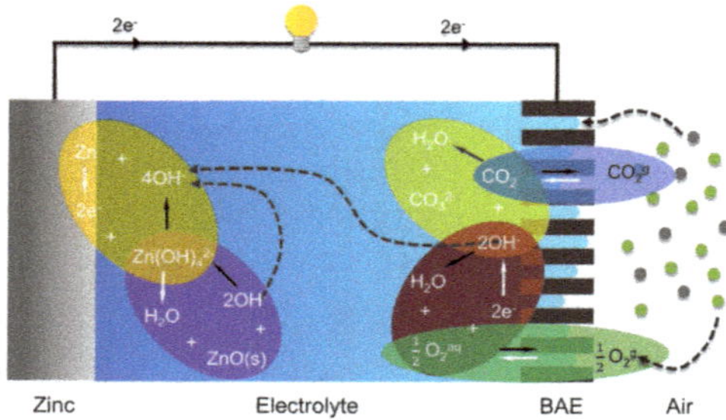

Figure 2. Operational schematic of an alkaline zinc-air battery. The various (electro)chemical reactions are indicated by the colored ovals; white arrows indicated discharging and black arrows indicate charging. Dashed lines show important transport paths.

2.2. Challenges, Progress, and Opportunities

While ZABs have been quite successful as primary cells, there are a number of hurdles that limit their electrical rechargeability and provide opportunities for further research.

The most well-known challenges relate to the aqueous KOH electrolyte. When the ZAB is operated in air, CO_2 can dissolve in the electrolyte and react with OH^- to form $CO_3{}^{2-}$ [56]. This parasitic reaction reduces the conductivity of the electrolyte, slows down the cell reactions, and eventually kills the cell [55]. As such, the lifetime of alkaline ZABs is limited from the moment they are exposed to air. A CO_2 filter could be applied to scrub the air [57], but this adds cost and complexity to the system. A second challenge for the electrolyte is the evolution of hydrogen gas. The potential of the Zn electrode reaction is below the potential for hydrogen evolution, which causes the electrolyte to be thermodynamically unstable [58]. However, H_2 evolution on the Zn surface can be kinetically suppressed with dopants, such as Hg, In, or Bi [19,59–61].

To address these challenges, current research is focused on the development of alternative electrolytes. An ideal electrolyte for Zn-air batteries should be stable in the electrochemical window of the cell, stable in air, conductive, non-corrosive, and thermodynamically favor the reversible precipitation of the desired final discharge product. Alternatives include aqueous alkaline with additives [62], which were recently evaluated by Schröder et al. [63] and Mainar, et al. [64,65]. Near-neutral chloride-based electrolytes have been experimentally evaluated by Zong and co-workers [43,44] with promising initial results. Chloride-based electrolytes address the carbonation issue and could improve the quality of Zn deposition, but they struggle with the precipitation of zinc-chlorides and the strongly oxidative nature of chlorine. Non-aqueous ionic liquid electrolytes [45,46,66–68] have been proposed with promising results [69], but the rate of these systems is limited and much work remains to be done.

Metal electrodes offer the possibility of achieving very high energy density. However, the formation of mossy or dendritic metal structures during charging can cause the electrode to change shape [70] and lead to an internal short-circuit [71], killing the cell. With its low surface diffusion characteristics and fast deposition kinetics [72], Zn is specifically vulnerable to electrode shape change. Achieving homogeneous Zn deposition is essential to the development of a secondary ZAB. Electrolyte additives and alternative electrode designs have been proposed to address this challenge [43,73–77], with mixed results.

A further challenge in Zn electrode design is the passivation of the electrode surface due to ZnO precipitation [78]. To achieve a high energy density, the precipitation of ZnO is required. However, when ZnO precipitates on the active surface of the Zn metal, it limits the transport of species to/from the electrolyte. When this ZnO barrier becomes too large, the electrochemical reaction can no longer proceed and the electrode is passivated [62,79]. ZnO passivation can take on two forms [54,80]. Type I ZnO has a porous white morphology. It precipitates near the surface of the Zn particle, but does not block it completely. The formation of type I ZnO is reversible. Type II ZnO forms a black dense layer directly on the surface of the the Zn particle. It is thought that type II ZnO forms on the active sites of the Zn metal, permanently blocking them and creating a passivating avalanche effect [81].

Finally, the development of an active, stable, and cheap bi-functional air catalyst is a hotly pursued topic in material science [44,51,82–84]. It is difficult to find a catalyst that is suitable for both the ORR and the OER. The alternating oxidative and reductive environments present in the BAE during cycling further complicates this challenge, and tends to accelerate the degradation of both the BAE carbon substrate and the catalyst. This difficultly is compounded for near-neutral electrolytes, in which the pH of the electrolyte may vary within the buffering range. From a material science perspective, research into new catalyst combinations and non-carbon BAE substrates is on-going. Alternative 3-electrode cell designs have also been proposed [19].

On a system level, the challenges described above have led to creative engineering solutions including mechanically rechargeable Zn-air fuel cells with some niche applications [69]. However, the logistical challenges of these designs prevented them from being adopted on a wide scale. The goal is to develop a high-performance electrically rechargeable Zn-air battery. To achieve this, new ideas and novel designs are needed on every level from material science up through system engineering. In the following section we discuss how numerical modeling and simulation can help in this pursuit and give an overview of recent progress.

3. Numerical Modeling and Simulation

Experimentally-based research can be expensive and time consuming. The development of theory-based models can help guide researchers down the most promising paths, provide a framework for interpreting experimental results, and lead to new breakthroughs in battery design.

Numerical modeling and simulation serve many purposes in metal-air battery research and can span a wide range of space and time-scales [30,85,86]. In the following sections, we review modeling methods commonly used in development of metal-air batteries. We begin on the material level, highlighting methods for studying the electronic properties of catalysts and metal electrodes as well as electrolyte thermodynamics and composition. We then move up to the electrode level, and discuss numerical studies of electrode architecture and what considerations are important in the design process. Finally, we give an overview of cell-level continuum modeling and discuss recent contributions to the literature.

3.1. Material Development

The first step in designing a feasible battery must be the selection of appropriate materials. Anode, cathode, and electrolytes all have their own, sometimes conflicting, requirements for activity and stability. Applied modeling methods can help screen potential materials, identify promising paths for development, and reduce the reliance on trial-and-error approaches.

3.1.1. Electrode Materials

Catalyst development is one of the most expensive and time consuming aspects of material research in metal-air systems. Bi-functional air catalysts can take on a wide range of compositions, often include expensive or toxic metals (e.g., Pt, Ag, Co, etc.), and are labor intensive to produce and test. Atomic-scale modeling methods can be applied early in the design process to screen potential catalysts.

Density functional theory (DFT) uses quantum mechanical calculations to make predictions about the electronic structure of multi-atom systems [87]. This method allows researchers to investigate the properties of materials considering the influence of things like surface structures and local coordination of atoms. The basic approach of DFT is to analyze a multi-atom system as the movement of electrons through a fixed array of atomic nuclei. Using the Born-Oppenheimer approximation, the state of the nuclei and the electrons can be split into separate mathematical expressions [88]. In this way the adiabatic potential energy surface of the atoms can be calculated and used to investigate the characteristics of the material [87]. Due to the complexity of the calculations involved, the method is confined to considering a limited number of atoms. Nonetheless it has been shown to be effective at screening the properties of metal alloys for a variety of applications [89].

Beginning with the work of Norskov, et al. [90], there have been a wealth of DFT studies investigating the reaction pathway, activity, and stability of materials in metal-air systems. Viswanathan et al. have utilized DFT to predict the activity of different pure and alloyed metals [91,92]. This group has also applied DFT calculations to investigate a range of phenomena in Li-air batteries [93,94]. In 2010, Keith and Jacob clarified a multi-pathway electrochemical mechanism for the ORR, which showed good agreement with experimental data. Eberle and Horstmann correlated the change in reaction pathway to an observed change in the Tafel slope [95]. In 2017, Tripkovic and Vegge elucidated the mechanism of the ORR on Pt (111), and used their results to investigate the high activity of Pt-alloys [96]. Non-metallic catalysts, such as nitrogen-doped graphene, are desirable due to their safety and low cost. A multi-scale model featuring DFT of such systems was recently presented by Vazquez-Arenas, et al. [97], and used to investigate the rate-determining step for the ORR in KOH electrolyte

DFT simulations can also be applied to investigate metal electrode materials and solid precipitants. Siahrostami, et al., modeled the effect of surface structure on zinc dissolution [98]. They applied their model to simulate the dissolution of a Zn kink atom, highlighting the potential steps of the dissolution process and predicting the overpotential of the reaction. For electro-deposition, Jäckle and Groß propose that surface diffusion processes are key to understanding the formation of metallic surface structures [99]. They utilized a DFT model to evaluate a range of metal anode materials for their tendency to form dendrites. In the case of non-aqueous Li-air batteries, the precipitation of Li_2O_2 can electrically isolate the cathode. DFT simulations have been applied to investigate the growth and electronic structure of Li_2O_2 [31,100–103], in an attempt to mitigate the risk of passivation.

3.1.2. Electrolytes

The behavior of liquid electrolytes in electrochemical systems can be quite complex and have a deciding influence on overall cell performance. The first step in determining the suitability of an electrolyte for a given system is to examine its equilibrium thermodynamics.

The speciation of ions, solubility of solids, and equilibrium potential of electrodes in an electrolyte is strongly dependent on pH and solute concentration [80,104,105]. For some systems, such as $KOH-ZnO$, this behavior is rather straightforward and well-documented [80,106], while for others, such as $ZnCl_2-NH_4Cl$, it is very complex and sensitive [3,107]. Figure 3a shows the speciation of the Zn^{2+} ion in $ZnCl_2-NH_4Cl$. Understanding this behavior can be helpful in interpreting battery performance and optimizing system design.

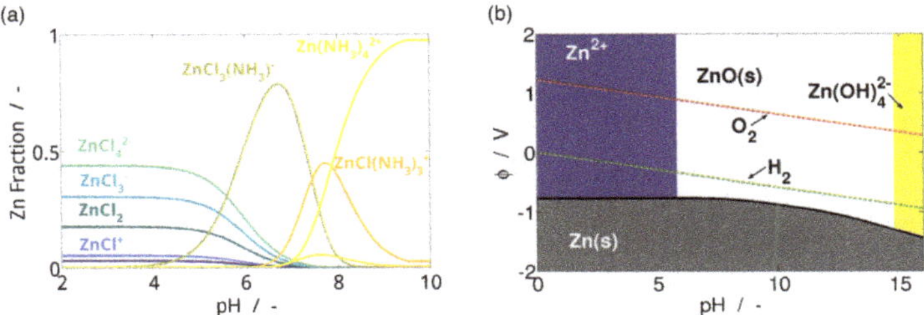

Figure 3. (**a**) Zn speciation plot for 0.51 M $ZnCl_2-2.34MNH_4Cl$ with pH adjusted by NH_4OH reproduced from Clark, et al. [3] with permission from Wiley-VCH and (**b**) simplified Pourbaix diagram for $[Zn]_T = 0.5$ M.

According to the law of mass action, for a system at equilibrium with a constant temperature, the value of the reaction quotient is constant. For a generic reaction, this concept is expressed as,

$$aA + bB \rightleftharpoons cC + dD, \qquad \frac{[A]^a[B]^b}{[C]^c[D]^d} = \beta. \tag{5}$$

The value of the constant, β, is referred to as the thermodynamic stability constant (sometimes also called the formation or equilibrium constant). In the 1970s, Smith and co-workers assembled an exhaustive compilation of thermodynamic stability constants of metal-ligand complexes, pKa values of acids, and solubility products of solids for a myriad of both inorganic and organic molecules [108–111]. By incorporating these equilibrium expressions for every electrolyte reaction into simple equations for the conservation of mass and charge, the equilibrium composition of the electrolyte and the solubility of solids can be predicted as a function of pH and solute concentration. A framework for such a model [112,113] was presented by Limpo, et al., in the 1990s, and was expanded upon in more recent research [106,114]. Clark, et al., recently presented a framework for predicting the discharge characteristics of a Zn-air cell with an aqueous near-neutral electrolyte based on equilibrium thermodynamic considerations [3], and showed how they can be incorporated into a dynamic model.

Models of equilibrium thermodynamics can also be used to generate potential-pH, or Pourbaix, diagrams [115]. Figure 3b shows a simplified example for the aqueous Zn system. The Nernst equation describes the connection between solute concentration and the equilibrium electrode potential [116]. The equilibrium potentials for the electrochemical reactions in Zn-air systems [80] can be expressed as:

$$E_{Zn/Zn^{2+}} = -0.762 + \frac{RT}{2F}\ln([Zn^{2+}]), \tag{6}$$

$$E_{H_2/H_2O} = 0 - 2.303\frac{RT}{F}\,pH, \tag{7}$$

$$E_{O_2/H_2O} = 1.229 - 2.303\frac{RT}{F}\,pH. \tag{8}$$

Taking this into consideration, researchers can predict the influence of shifts in pH or solute concentration on electrode potential. Aside from describing the voltage of the cell, this is also useful for identifying possible parasitic reactions, such as H_2 evolution or redox shuttles.

In addition to thermodynamic models, DFT can be used to screen properties of electrolytes and their suitability for metal-air applications. One of the major challenges in aprotic LABs is the development of a solvent that is stable and facilitates oxygen solubility and transport. In 2015, Husch and Korth presented a study of non-aqueous LAB electrolytes [117]. In a wide-ranging work requiring about 2 million hours of process time, they integrated DFT calculations into a larger framework to screen 927,000 potential electrolyte solvents for high Li^+ and O_2 solubilities and low viscosity. An illustration of their work is shown in Figure 4. By strategically applying modeling methods, Husch, et al., were able to by pass the trial-and-error approach and directly highlight electrolyte solvents with the highest chances of success.

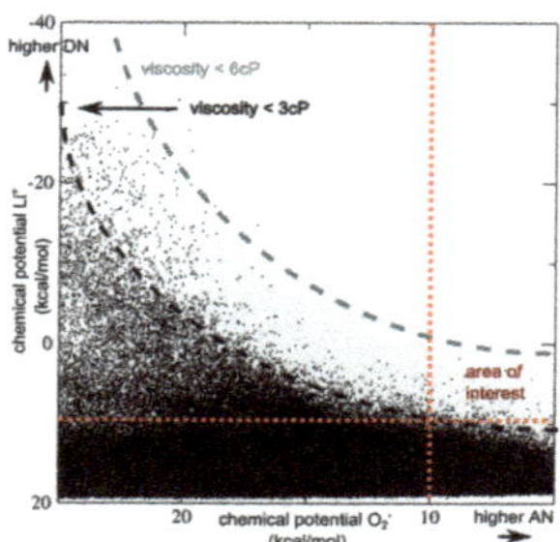

Figure 4. Chemical potentials for Li^+ vs. chemical potentials for O_2 in the bulk candidate compound are plotted for all 927,000 compounds. Black dots indicate compounds with a viscosity below 3 cP, grey dots indicate compounds with a viscosity below 6 cP. Reproduced from Husch, et al. [117]—Published by the Physical Chemistry Chemical Physics (PCCP) Owner Societies.

3.2. *Electrode Design*

The development of active and stable materials is essential to battery design. However, care must be taken to appropriately scale-up materials to the electrode level. In this section, we discuss modeling tools which can help fashion suitable materials into high-performance electrodes.

3.2.1. Bi-Functional Air Electrode (BAE)

The BAE in Zn-air batteries is comprised of a porous structure (usually carbon fibers and binder) with mixed hydrophilic and hydrophobic properties. The idea is to encourage the formation of the so-called three-phase boundary without either over-saturating (flooding) or under-saturating (drying out) the pores. When the cell is cycled, pressure gradients are induced in the battery due to the precipitation and dissolution of solid products, which can change the saturation of the BAE. To simulate this behavior, a method of predicting the pressure-saturation characteristics of the BAE structure is needed.

The Lattice Boltzmann Method (LBM) is useful for simulating multi-phase flow in porous media. This approach uses the Boltzmann equation to simulate the flow of fluids as a combination of collision and streaming events of particles on a discrete lattice [118,119]. Particle positions are confined to the nodes of the lattice and it is assumed that they can move between their current position and adjacent nodes in discrete lattice directions. The probability to find particles at a lattice node with a velocity component in any of the discrete directions is described by a distribution function [120]. If solid boundaries are present in the system, no-slip boundary conditions can be introduced by a simple bounce back scheme. Implementation of the basic LBM equations is straightforward, and there are a variety of open-source codes available [121].

In electrochemical research [122], Lattice Boltzmann models have been applied to investigate the transport of water in the GDL of PEM fuel cells [123–126]. One significant benefit of LBM is the ability to simulate flow in complex geometries. Using X-ray tomography [127] or focus ion beam scanning electron microscope (FIB-SEM) images [128], real electrode structures can be modeled in 3D and their transport properties evaluated. Recently, Danner, et al., presented a LBM model to predict the pressure-saturation parameters of BAEs in metal-air batteries [128]. Figure 5 shows pressure-saturation curves calculated with LBM for both 2D and 3D simulations of real BAE structures. Their simulations show that the pressure-saturation characteristics of air electrode substrates vary according to whether the electrolyte is draining from the structure (configuration I) or imbibing the structure (configuration

II). They proposed this is because the structure contains some pores that can be filled with electrolyte, but are not easily emptied.

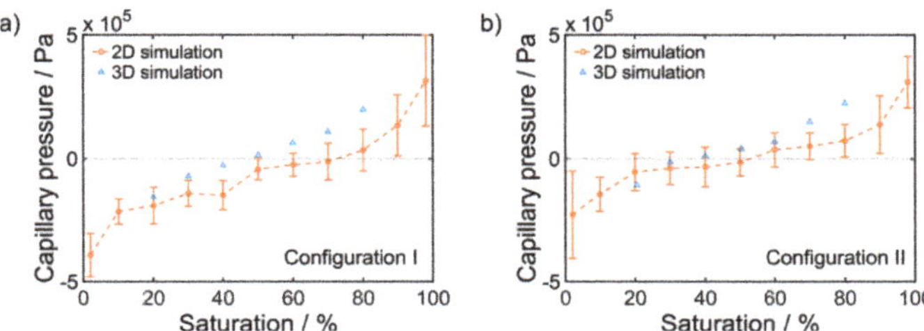

Figure 5. Pressure-saturation curves for real air electrode structures simulated with 2D and 3D Lattice-Boltzmann models for (**a**) draining (configuration I) and (**b**) imbibition (configuration II). Reproduced from Journal of Power Sources, 324, T. Danner, S. Eswara, V.P. Schulz, A. Latz, Characterization of gas diffusion electrodes for metal-air batteries, 646–656, Copyright 2016, with permission from Elsevier [128].

3.2.2. Metal Electrode

The design of the metal electrode is determined mostly by effects linked to passivation and shape change. For Zn-air batteries, the precipitation of ZnO on the electrode surface can isolate it from the electrolyte and slow down the reaction kinetics, eventually killing the electrode. It has been shown that ZnO can take on a porous white morphology (type I) that is reversible, or a dense black morphology (type II) that is irreversible. To model the effects of these precipitants on electrode performance, various models have been developed.

As discussed in Section 2.2, type I ZnO is formed when the dissolved zinc concentration in the electrolyte exceeds the saturation limit, and it precipitates near the electrode surface. This layer of porous ZnO is generally modeled as an additional mass transport barrier [55], slowing diffusion and migration of OH^- to the electrode surface. Early models determined the passivation characteristics using the so-called Sand equation, an empirical expression linking current density, i, and passivation time, t, with constants, k and i_e:

$$i = kt^{0.5} + i_e. \tag{9}$$

In 1981, Liu, et al., expanded this concept taking into account the mechanism for type I ZnO precipitation. They proposed that the passivation due to type I ZnO occurs via a dissolution-saturation-precipitation mechanism. Put simply, the Zn electrode dissolves until the concentration of $Zn(OH)_4{}^{2-}$ exceeds the saturation limit for nucleation, and the ZnO phase grows as the precipitation reaction proceeds. In their model, they calculate the time required to saturate the electrolyte with $Zn(OH)_4{}^{2-}$ (t_a), the time to precipitate type I ZnO (t_b), and the time to precipitate type II ZnO (t_c), and define the passivation time as $t = t_a + t_b + t_c$ [78]. The resulting 0D model is a helpful predictor of Zn electrode performance, but is not suitable for use in continuum modeling.

In the continuum model of Zn electrodes developed by Sunu and Bennion in 1980 [129], they considered passivation by assuming that the precipitation of ZnO reduced the active surface area available for the Zn dissolution reaction. More recently in 2017, Stamm, et al., implemented the effect of type I ZnO passivation in a continuum model by calculating the thickness of the ZnO shell and numerically solving for the species concentration at the surface [55], assuming Nernst-Planck transport across the barrier. These values were then used to calculate the Nernst potential and exchange current density of the Zn dissolution reaction.

While models for passivation due to type I ZnO are rather well developed, there are fewer models for type II ZnO passivation. In 1991, Prentice, et al. [81] proposed that type II ZnO forms directly on the surface of the Zn electrode, and does not follow the dissolution-saturation-precipitation mechanism of type I. By calculating the fractional surface coverage of various zinc-hydroxides as a function of concentration and electrode potential, they were able to simulate rotating disk experiments. Their simulations agreed well with experimental measurements. The model was recreated and the results are shown in Figure 6.

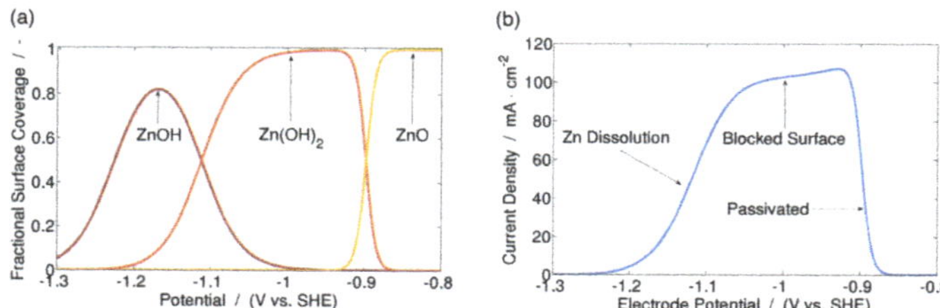

Figure 6. (**a**) Fractional surface coverage of species on the Zn electrode surface, and (**b**) simulated LSV measurement of type II ZnO passivation. Recreated from the model described by Prentice, et al. [81].

While much progress has been made in understanding the mechanisms of Zn passivation, work remains to be done. A more unified theory of type I and type II passivation, along with implementation in a continuum model, could be an area for future research. A dedicated review of experimental and modeling studies of Zn electrode passivation was presented by Bockelman, et al., in 2017 [130].

3.3. Cell Modeling

The modeling methods described above are very useful for evaluating the properties of individual materials or components, but researchers often need to know how these components will interact with each other in a real electrochemical cell.

Continuum models are among the most useful and widely-applied methods for studying the cell-level performance of metal-air batteries. This approach applies the mass and charge continuity equations to describe the transient characteristics of spatially discretized systems [85,131]. A list of important continuum models for metal-air systems from the literature is presented in Table 1. For most applications, a 1D model is sufficient to describe the system. However, 2D and 3D [132,133] finite volume models of batteries can give more in-depth information for detailed analysis. The mass and charge continuity equations can be expressed in generic terms as,

$$\text{Mass Continuity:} \quad \frac{\partial(c_i\varepsilon_e)}{\partial t} = \underbrace{-\vec{\nabla}\cdot\vec{N}_i^{D,M} - \vec{\nabla}\cdot\vec{N}_i^{C}}_{\text{transport}} + \overbrace{\dot{s}_i}^{\text{source}}, \tag{10}$$

$$\text{Charge Continuity:} \quad 0 = \underbrace{-\vec{\nabla}\cdot\vec{j}}_{\text{transport}} + \overbrace{\sum_i z_i\dot{s}_i}^{\text{source}}, \tag{11}$$

where c_i is the concentration of solute i, ε_e is the electrolyte volume fraction, $\vec{N}_i^{D,M}$ is the diffusion-migration flux, $\vec{N}_i^{C}$ is the convective flux, $\dot{s}_i$ is the reaction source term, $\vec{j}$ is the electrolyte current density and z_i is the solute charge number. A detailed derivation of these terms based on non-equilibrium thermodynamics and their applicability to metal-air systems can be found in existing works [3,30,55,134]. In their most general form, the continuity equations describe the local conservation of mass and charge due to transport across the boundaries of a control volume and the presence of a source/sink within the bulk of the control volume. To successfully implement these equations physical models for electrolyte transport and (electro)chemical reactions are needed.

Table 1. Comparison of continuum models for metal-air systems. Transport models listed are concentrated solution theory (CST) or dilute solution theory (DST).

Year	Authors	System	Dimension	Transport Model	Notes	Source
1980	Sunu, et al.	Zn-Air	1D	CST	Zn & ZnO shape change	[129]
1991	Mao, et al.	Zn-Air	1D	CST	Precipitation of $K_2Zn(OH)_4$	[135]
2002	Deiss, et al.	Zn-Air	1D	DST	Concentration profiles and cell voltage	[136]
2010	Andrei, et al.	Li-Air	1D	CST	LAB modeling framework	[137]
2011	Albertus, et al.	Li-Air	1D	CST	O_2 transport and Li_2O_2 precipitation	[26]
2012	Neidhardt, et al.	Multiple	1D	CST + Multi-Phase	Flexible framework, convective transport	[131]
2013	Horstmann, et al.	Li-Air	1D	CST + Multi-Phase	Inhomogeneous Li_2O_2 precipitation, aqueous electrolyte	[35]
2014	Danner, et al.	Li-Air	1D	CST + Multi-Phase	Air electrode model with pressure-saturation	[36]
2014	Schröder, et al.	Zn-Air	0D	CST	Effect of air composition on cell performance	[138]
2014	Arlt, et al.	Zn-Air	0D	CST	State-of-charge monitoring with x-ray CT	[139]
2014	Xue, et al.	Li-Air	1D	CST	Li_2O_2 pore clogging with pore size distribution	[140]
2015	Grübl, et al.	Li-Air	1D	CST + Multi-Phase	Engineering evaluation of system design	[141]
2016	Yin, et al.	Li-Air	1D	DST	Affect of Li_2O_2 particle size on charging profile	[142]
2017	Stamm, et al.	Zn-Air	1D	CST + Multi-Phase	Affect of ZnO nucleation and growth on cell discharge profile	[55]
2017	Clark, et al.	Zn-Air	1D	CST + Multi-Phase	Framework for buffered near-neutral electrolytes	[3]

Chemical reaction models feature a term describing the thermodynamic driving force and an expression of the kinetics [143]. In the case of electrochemical reactions, the most widely-used model is the Butler-Volmer approximation [144],

$$k = k_0\left(\exp\left[\frac{\alpha RT}{nF}\eta\right] - \exp\left[-\frac{(1-\alpha)RT}{nF}\eta\right]\right), \tag{12}$$

where k_0 is the rate constant (linked to the exchange current density), α is the symmetry factor, η is the surface overpotential, and the other variables take on their usual meaning. While this approximation is sufficient to describe simple electrodes, the precipitation of solid metal-oxides on the surface of the metal electrode forms an insulating layer and can cause the kinetics of the electrode to deviate from idealized models [62,79,145]. Special models of metal-electrode kinetics considering the effects of passivation have been developed [78,81,130] and implemented [55] in continuum simulations.

The Marcus theory of charge transfer reaction kinetics is a more accurate alternative to the Butler-Volmer approximation [146–148]. The Marcus model builds on an Arrhenius approach, in that the pre-exponential factor is described by the electronic coupling element, H_{ab}, and the reorganization free energy, λ, and the exponential term containing the activation energy. In its quantum mechanical form, the Marcus theory is expressed as

$$k_{ct} = \frac{2\pi}{\hbar}\frac{\mid H_{ab}\mid^2}{\sqrt{4\pi k_B T\lambda}}\exp\left[-\frac{(\lambda+\Delta G^0)^2}{4k_B T\lambda}\right]. \tag{13}$$

Marcus theory results naturally from quantum mechanics, and can be more easily linked to simulations like DFT [149]. While this approach has been applied in some continuum models, it is difficult to parameterize.

Electrolyte transport is modeled using a combination of expressions for diffusion, migration and convective mass flux [116], as well as a source term stemming from the chemical reactions described above [150,151]. While the fundamental components of electrolyte transport models are universal, their exact form can vary based on the ionic strength [116] and pH [3] of the electrolyte. For low ionic strength electrolytes, a simplified dilute solution theory (DST) approach can be applied to model the diffusion and migration transport of solutes [116].

$$\text{Dilute Solution Theory:} \quad \vec{N}_i^D = -D_i\vec{\nabla}c_i, \qquad \vec{N}_i^M = \tfrac{D_i c_i z_i F}{RT}\vec{\nabla}\phi_e, \tag{14}$$

$$\text{Concentrated Solution Theory:} \quad \vec{N}_i^{D,M} = -D_i\vec{\nabla}c_i - \tfrac{t_i}{z_i F}\vec{j}, \qquad \vec{j} = -\kappa\vec{\nabla}\phi_e + \tfrac{\kappa}{F}\textstyle\sum_i \tfrac{t_i}{z_i}\tfrac{\partial\mu_i}{\partial c_i}\vec{\nabla}c_i. \tag{15}$$

For high ionic strength electrolytes, a more complete concentrated solution theory (CST) is needed. In this case, a coupled expression for diffusion-migration flux can be derived from non-equilibrium thermodynamics [134,152,153]. For strongly acidic or alkaline electrolytes, the concentrations of H^+ or OH^- are usually so high that concentration gradients do not affect the thermodynamic stability of the solutes. However, for electrolytes in the weakly acidic to weakly alkaline range, the concentrations of solutes in electrolytes can swing by orders of magnitude as pH gradients develop in the cell [80,112,113] and can affect the performance of the cell [3]. A new method for modeling electrolyte transport in near-neutral systems was recently proposed by Clark, et al. [3].

When metal-air batteries are operated, the precipitation and dissolution of solids and the conversion of H_2O by the ORR/OER can induce a convective flux of electrolyte in the cell [30]. In general terms, the convective flux can be expressed as

$$\vec{N}_i^{\,C} = c_i\vec{v}_e, \tag{16}$$

where $\vec{v}_e$ is the center-of-mass velocity of the electrolyte. This adds an additional level of complexity to metal-air battery models over closed systems, such as Li-ion, which often apply a simpler Nernst-Planck model. A method for considering multi-phase convective flow in continuum models was presented by Horstmann, et al. [30] and is discussed in detail later in the text.

Continuum models of ZABs have been developed intermittently since the 1980s. The first 1D continuum model of a Zn electrode in an alkaline ZAB was developed in 1980 by Sunu and Bennion [129]. It was based on the general 1D model for concentrated transport in porous electrodes outlined by Newman [116]. Their simulations showed the inhomogeneous precipitation of ZnO and investigated the shape change of the Zn electrode during cycling. In 1992, Mao and White [135] developed an extended model that also resolved the separator and air electrode. They found that $K_2Zn(OH)_4$ does not precipitate and compared simulated cell voltages with experimental measurements. Ten years later, Deiss, et al. [136] performed ZAB cycling simulations with a 1D model of the Zn electrode and separator based on dilute solution theory. They studied the redistribution of Zn and the development of concentration gradients in the cell.

In recent years, there has been a boom in continuum modeling frameworks for both Zn-air and Li-air systems, with some areas of overlap. In 2012, Neidhardt, et al., presented a flexible continuum modeling framework for multi-phase management, with direct application to a variety of electrochemical systems [131]. In their work, they applied this framework to simulate a range of fuel cells and batteries to demonstrate the versatility of the approach. In their simulations of non-aqueous $Li{-}O_2$ batteries, they noted that the system is limited by a combination of slow oxygen transport and blockage of cathode pores with Li_2O_2.

To address the oxygen transport and passivation challenges associated with aprotic LABs, Horstmann, et al., were motivated to examine precipitation in alkaline aqueous LABs [30]. Their model featured two important developments. The first was the introduction of pressure-saturation expression to simulate the electrolyte flooding and drying-out of the BAE. The saturation of porous structures was described with a so-called Leverett approach, which had previously been applied in models of fuel cells. The Leverett function, $J(s)$, uses empirical constants to approximate the saturation of a porous structure as a function of the capillary pressure, $(p_e - p_g)$:

$$J(s) = \sqrt{\frac{B_e}{\varepsilon_0 \sigma^2}}(p_e - p_g) = A + Be^{C(s-0.5)} - De^{-E(s-0.5)}. \tag{17}$$

In this expression, B_e is the electrolyte permeability, ε_0 is the porosity of the electrode, σ is the surface tension, and p_e and p_g are the pressure in the electrolyte and gas phases, respectively. In the expression for the Leverett function, s is the electrode saturation and the remaining variables are constants. These constants can be determined experimentally or predicted numerically by combining 3D structure characterization with LBM simulations [128]. With the pressure of the electrolyte and the saturation of the BAE known, the convective velocity of the electrolyte can be solved using a Darcy model:

$$\vec{v}_e = -\frac{B_e}{\eta_e}\vec{\nabla} p_e, \tag{18}$$

where η_e is the viscosity of the electrolyte. Their simulations found that the availability of gas diffusion electrodes for aqueous systems reduces the oxygen transport limitations seen in non-aqueous LAB systems.

The second development was the implementation of a model for $LiOH \cdot H_2O$ precipitation based on the classical theory of nucleation and growth. By defining terms for the reaction enthalpy of formation for both the bulk and surface of the nucleus, they identified the critical formation energy and nucleus size. Considering a diffusion limited precipitation mechanism and the supersaturation of Li^+ as the driving force for nucleation and growth, they were able to simulate the spatially resolved precipitation of $LiOH \cdot H_2O$. The results, shown in Figure 7, indicate that $LiOH \cdot H_2O$ does not block the cathode pores. Rather it precipitates mostly near the separator-anode interface, thereby addressing the passivation challenge in aprotic LABs.

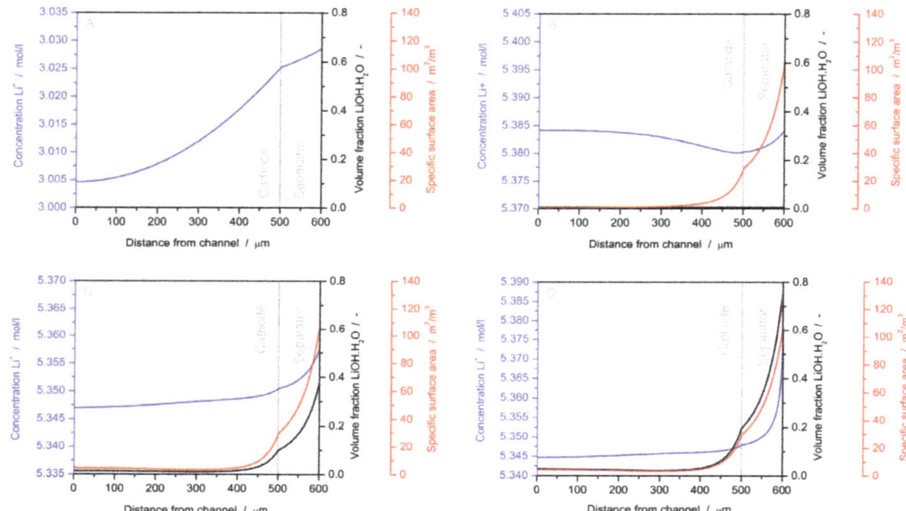

Figure 7. Spatial profiles of salt concentration, volume fraction of $LiOH \cdot H_2O$, and the specific surface are of precipitation during galvanostatic cell discharge (i = 10 $mA \cdot cm^{-2}$). Li^+ concentration increases before $LiOH \cdot H_2O$ nucleates (**A**) As the discharge progresses, $LiOH \cdot H_2O$ nucleates (**B**) and precipitates (**C**) until the cell fails due to a solid film forming at the separator-anode interface (**D**). Precipitation occurs mainly close to the anode due to the small transference number of Li^+. Reprinted from Horstmann, et al. [30]—Reproduced by permission of The Royal Society of Chemistry.

In 2015, the model of Horstmann, et al., was extended by Grübl and Bessler to engineer seven variants of aqueous alkaline LAB systems [141]. They identified improvements to the glass separator and the development of lighter electrode materials as areas for further research. While the potential advantages of aqueous LABs are clear, most recent modeling studies have focused on non-aqueous systems.

One of the first continuum models for non-aqueous LABs was presented by Andrei, et al., in 2010 [137]. Their simulations considered the effects of cell architecture and operational conditions on concentration profiles and cell voltage, and provided a solid foundation for further development. Recent multi-scale modeling studies of non-aqueous LABs focus heavily on the nucleation and growth of solids, and their affect on cell performance. A 1D continuum model of a LAB was developed by Albertus, et al., in 2011 [26]. They found that although O_2 transport can be limiting for high current densities, the main limitation in LABs relates to the precipitation of solids. For carbonate solvents, the dominant discharge product is Li_2CO_3, which, along with Li_2O_2, poses a strong passivation risk to the cathode. The model was based on a CST approach for electrolyte transport. It did not consider the effects of electrolyte convection or cathode saturation, which was identified as an area for future development.

With the shift to non-carbonate electrolytes, the morphology of Li_2O_2 precipitated during discharge became the subject of research. Knowing that pore blockage was a challenge in LAB performance, Xue, et al., developed a continuum model to investigate Li_2O_2 precipitation considering the pore size distributions of cathode materials [140]. They later extended their model to determine the effect of both electrolyte solvent and applied current density on Li_2O_2 morphology [154]. A nano-scale continuum model to study the rate-dependent growth of Li_2O_2 was presented by Horstmann, et al. [35]. They found that the morphology of Li_2O_2 shifts from discrete particles at low discharge rates to an electronically insulating film at high current densities. This line of research was then expanded to consider the effects of Li_2O_2 precipitation on the charging process. Yin, et al., presented a continuum model for non-aqueous LABs that links the size of the Li_2O_2 particles created during discharging to the two-step voltage profile observed during charging [142]. The effect of the Li_2O_2 reaction mechanism on the discharge/charge characteristics was investigated by Grübl, et al. [155]. They found that the reaction mechanism is partially irreversible, and considered the effects of adding a redox mediator to the electrolyte. Finally, some researchers have highlighted not only the effects of Li_2O_2 morphology, but also its electronic properties. Radin, et al. integrated a DFT simulation of Li_2O_2 with charge carrying dopants into a simple Nernst-Planck continuum model to study ways to promote the OER [103]. They found that dopants, such as Co and Ni could enhance the OER and improve the rechargeability of non-aqueous LABs.

Applying continuum modeling methods to both aqueous and non-aqueous LAB development has illuminated the challenges and the opportunities inherent to these systems. Further research is needed into the effects of solid precipitation and oxygen transport in non-aqueous LABs and the long-term electrolyte stability of aqueous LABs. However, the modeling studies highlighted above have shown promising paths for further investigation.

Zn-air continuum modeling studies provide insight into challenges, such as electrolyte carbonation, Zn electrode passivation, and improved cell design. In 2014, Schröder, et al. published a framework for a 0D ZAB continuum model, which they utilized to study the effect of air composition on cell performance [138]. Examining the effects of the relative humidity (RH) and carbon dioxide content of air, they found that controlling the RH can help reduce electrolyte loss and that the presence of CO_2 can dramatically limit the lifetime of the cell. In a separate paper, they combined this model with x-ray tomography measurements of a primary button cell to monitor the state-of-charge during discharge [139].

Experimental tests of ZAB button cells have shown a voltage step in the middle of discharge, particularly at high current densities [55]. In 2017, Stamm, et al., presented a model to clarify the mechanism behind this observation. Concentration profiles from their model are shown in Figure 8. The nucleation of ZnO requires an over-saturation of $Zn(OH)_4{}^{2-}$ in the electrolyte. For high current densities, the electrolyte concentration gradients that develop in the cell are strong enough that $Zn(OH)_4{}^{2-}$ does not reach the critical super-saturation for nucleation in the anode-separator interface and ZnO does not nucleate. As a result, the surface concentration of OH^- in this region is much higher than in areas of the electrode covered by a ZnO film, as shown in Figure 8c. When the uninhibited Zn near the separator is completely utilized, the overpotential of the dissolution reaction increases, causing the observed drop in cell voltage. For this reason, they proposed that Zn electrodes should contain a small amount of ZnO powder. In this way, the effects of inhomogeneous ZnO nucleation can be avoided. Stamm, et al., also considered the effects of CO_2 dissolution in the KOH electrolyte. They found that after about 2 months, the carbonation of the electrolyte becomes so severe that the cell can no longer function. To address this issue, they purpose employing either carbon dioxide filters or neutral electrolytes.

ZABs with near-neutral chloride-based electrolytes could address the electrolyte carbonation issue and have been experimentally investigated [43,44]. The initial results are promising, but the composition and behavior of these electrolytes during cell operation is unclear. In 2017, Clark, et al. presented a continuum framework for modeling pH buffered aqueous electrolytes, and applied it to study ZABs with pH adjusted $ZnCl_2-NH_4Cl$ electrolytes [3]. Utilizing a 0D thermodynamic model of the electrolyte, they determined the pH stability and predicted the conditions under which a range of solids would precipitate. Integrating this method into a 1D continuum model, they simulated the performance of experimental near-neutral ZABs from the literature. Figure 9 shows concentration profiles of in the cell proposed by Goh, et al. [43] during cycling. The Zn electrode is on the left and the BAE is on the right of the domain. To maintain a neutral pH in the BAE, the buffer reaction $NH_4{}^+ \rightleftharpoons NH_3 + H^+$ counteracts the pH shifts inherent in the ORR/OER. As more NH_3 is produced, it forms dominant complexes with Zn^{2+}, shown in Figure 9a. Because there is an excess of NH_4^+ in the electrolyte, the buffer reaction is uninhibited and the pH during discharging is relatively stable (Figure 9e). During charging, the buffer reaction is reversed and NH_3 is converted to NH_4^+. As NH_3 is depleted, zinc-chloride complexes dominate in the BAE (Figure 9b). Some of the NH_3 that was produced during discharge diffuses into the bulk electrolyte and cannot be quickly recovered. When NH_3 is locally depleted, the buffer reaction becomes limited and the pH in the BAE becomes acidic (Figure 9f). Acidic pH values can accelerate catalyst degradation and limit the lifetime of the cell. In their work, Clark, et al., discuss how cell architecture and electrolyte composition can be optimized to avoid this effect and improve performance.

Continuum models can be invaluable for investigating a range of phenomena in electrochemical cells, from the effects of discharge product precipitation to the electrolyte stability. Through the development of theory-based models, side-by-side with experimental investigation and validation, researchers can identify and pursue the most promising paths towards advanced metal-air batteries.

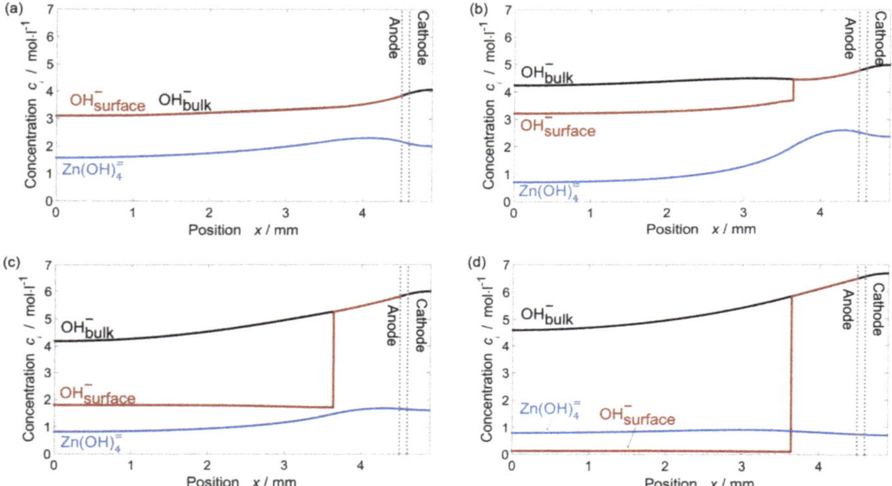

Figure 8. Various concentration profiles in an alkaline ZAB button cell during galvanostatic discharge at 125 Am^{-2} at different times. (**a**) Before ZnO nucleates, the OH^- concentration at the Zn surface and electrolyte bulk are equal and zincate concentration reaches its maximum; (**b**) ZnO precipitates inhomogeneously in the Zn-electrode, causing the OH^- surface concentration to be higher near the separator and fall as the ZnO barrier becomes thicker; (**c**) OH^- concentration continues to fall as ZnO passivation barrier grows; (**d**) OH^- concentration at the Zn-electrode surface is small and limits the further dissolution of Zn. Reproduced from Journal of Power Sources, 360, J. Stamm, A. Varzi, A. Latz, B. Horstmann, Modeling nucleation and growth of zinc oxide during discharge of primary zinc-air batteries, 136–149, Copyright 2017, with permission from Elsevier [55].

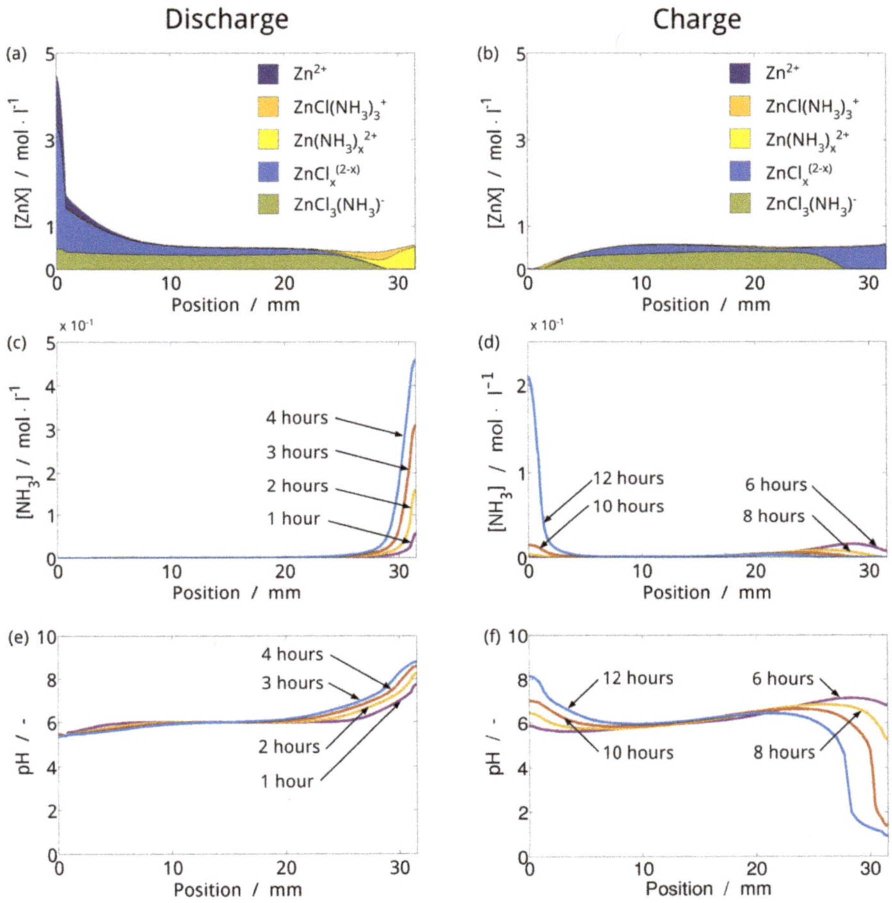

Figure 9. Electrolyte composition of near-neutral ZAB in during discharging and charging. At the end of discharging (**a**), zinc in the gas diffusion electrode (GDE) exists as $Zn(NH_3)_4^{2+}$. Once the capacity of zinc to take up NH_3 is completely utilized, NH_3 accumulates in the GDE (**c**); As the NH_4^+/NH_3 solution approaches its equivalence point, the pH value in the GDE becomes steadily more alkaline (**e**); At the Zn electrode, the small amount of NH_3 present is taken up by excess Zn^{2+} and the pH value becomes slightly more acidic. When the cell is charged, the production of H^+ in the GDE pushes the equilibrium of the ammonium buffer back towards NH_4^+. The zinc–ammine complexes release NH_3 back to the solution as charging progresses, and at the end of charging, zinc in the GDE exists exclusively as zinc–chloride complexes (**b**); To stabilize the pH value in the GDE, there must be NH_3 available for the conversion into NH_4^+. However, a considerable amount of the NH_3 produced during discharging diffuses into the bulk electrolyte and cannot be quickly recovered. This leads to a depletion of NH_3 in the GDE (**d**); At the Zn electrode, the concentration of NH_3 increases because of the redeposition of zinc. Without NH_3 to stabilize the pH value, the electrolyte in the GDE becomes acidic (**f**). At the Zn electrode, the loss of aqueous Zn^{2+} and the relative excess of NH_3 cause the pH value to increase. Reproduced from Clark, et al. [3] with permission from Wiley-VCH.

4. Model-Based Battery Engineering

To provide an example of how modeling and simulation can be applied to advance zinc-air battery development, we performed a series of cell-level continuum simulations using existing ZAB models [3,55]. In this analysis, we optimize the thickness of the zinc electrode to maximize the capacity of the cell while avoiding the unwanted effects of passivation. The initial composition of the electrolyte is 7 M KOH, saturated with ZnO. The BAE and separator are both 0.5 mm in length and the Zn-electrode is varied. The cell is galvanostatically discharged at current densities ranging from 0.1 to 50 $mA \cdot cm^{-2}$.

Figure 10a shows the magnitude of the KOH concentration drop across the cell at the end of charging. As the thickness of the Zn electrode and the magnitude of the current density increase, the long transport paths and large source terms induce significant concentration gradients in the cell. This is important because KOH gradients can affect the solubility of ZnO and increase the risk of passivation in the Zn electrode.

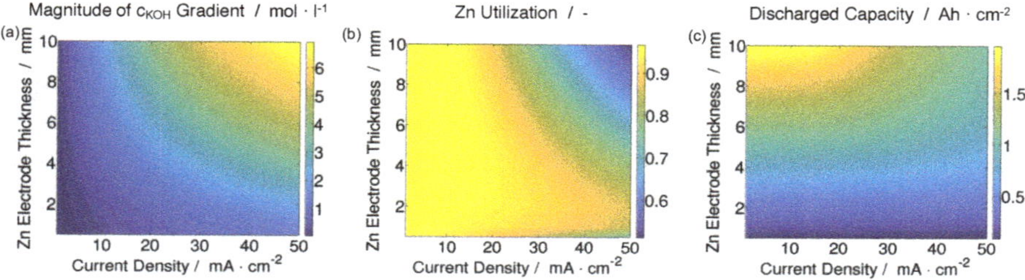

Figure 10. (**a**) Magnitude of the KOH concentration gradient across the cell at the end of discharge; (**b**) Zn utilization; and (**c**) discharged capacity as a function of Zn electrode thickness and current density.

Figure 10b presents the utilization of zinc metal in the battery. The results show that for current densities less than circa 20 $mA \cdot cm^{-2}$, the active Zn in the electrode is essentially completely utilized (>90%) for all Zn electrode thicknesses. However, for higher current densities, passivation of the electrode occurs due to two competing factors. The ZnO that precipitates in the electrode acts as a barrier to mass transport. For thick Zn electrodes, this barrier to transport is so large that the bulk concentration of KOH in the Zn electrode drops to the point that there is insufficient hydroxide present to form $Zn(OH)_4{}^{2-}$ complexes, and Zn utilization drops to circa 50%. For thin electrodes, the ZnO transport barrier remains relatively small and the bulk KOH concentration remains in an acceptable range. However, in these electrodes there is less active surface area available for the reaction, which leads to a higher flux term at the surface of the Zn particles. For current densities over 30 $mA \cdot cm^{-2}$ and electrodes less than 2 mm, the magnitude of the flux term is large enough to locally deplete OH^- at the electrode surface and passivate the electrode. The Zn utilization drops to circa 80%.

Figure 10c shows the discharged capacity of the battery. Increasing the thickness of the Zn electrode increases the amount of active material in the cell and the theoretically achievable capacity. However, the passivation of thick electrodes at higher current densities limits the amount of Zn that can be utilized, as shown in the previous figure. The result is that a battery with a 5 mm Zn electrode discharged at 10 $mA \cdot cm^{-2}$ has roughly the same capacity as a battery with a 10 mm Zn electrode discharged at 50 $mA \cdot cm^{-2}$.

When designing an alkaline ZAB, care should be taken to size the Zn electrode considering the current requirement and the desired capacity. With this information, an informed decision can be taken regarding how much Zn paste should be included in the battery to obtain the optimum performance.

Model-based engineering can also be applied in the testing phase of development to design experiments with the highest chance of success. Because Zn electrodes have a high capacity and can only be discharged at limited rates, it takes a very long time to study their cycling characteristics. Often, the solution is to perform "accelerated" tests with smaller electrodes or at higher current densities, both of which increase the risk of irreversible passivation.

Consider the case in which a researcher wants to cycle a ZAB 200 times in less than 3 months. One cycle is defined as moving between 70% and 30% state-of-charge (SOC), and the electrode consists of a paste that is 50 vol % Zn and 50 vol % electrolyte. Figure 11a shows the time required to complete 200 cycles under these conditions for different combinations of current density and Zn electrode thickness.

Figure 11b shows the time required to passivate the Zn electrode. By comparing the passivation time with the cycling time and applying an engineering safety factor of 1.3, we can mark the passivation limitations of the system. This region is shaded in black in Figure 11c. Combinations in the red zone of the figure exceed the time limitation, and we assume that electrodes smaller than 100 microns are impractical to manufacture (orange). With these factors in mind, we define a region of combinations (green) which could fit the researcher's needs.

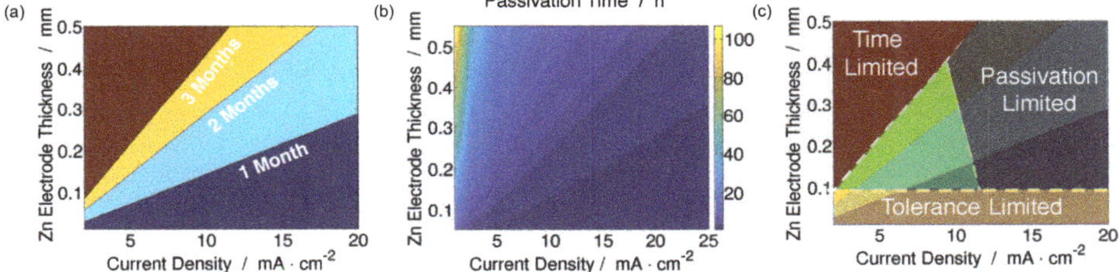

Figure 11. (**a**) Time required to cycle a ZAB 200 times for various Zn electrode thickness and current density combinations; (**b**) passivation time; and (**c**) operational window for a model ZAB.

When appropriately applied, model-based battery design can be of great value to scientists and engineers. The tools we have reviewed and applied in this analysis give insight into every aspect of battery performance, from the atomic structure of materials to the dynamic performance of whole cells. Embracing an integrated approach to modeling and understanding battery performance can help guide researchers towards achieving the goal of viable and high-performance metal-air batteries.

5. Conclusions

A variety of modeling and simulation methods can be applied to aid the development of zinc-air and other metal-air battery systems. While many metal-air systems are currently under development, Li-air batteries (LABs) and Zn-air batteries (ZABs) are the most promising systems.

On the material level, density functional theory (DFT) can be applied to investigate the electronic properties of catalysts and metals. This could help to screen new catalysts for properties like activity, stability, and selectivity and to elucidate the effect of surface structures on metal dissolution and deposition. Furthermore, equilibrium thermodynamic models can be used to predict the speciation of electrolytes and the solubility of precipitants. This can help determine not only the state of the electrolyte for different pH and concentration mixtures, but also its stability within the electrochemical window of the cell.

When it comes to electrode design, one challenge is to develop a bi-functional air electrode (BAE) that maintains an optimum level of saturation (neither flooding nor drying out) during battery operation. Lattice-Boltzmann-Methods (LBM) can be developed to investigate the pressure-saturation characteristics of real BAE structures in 2D or 3D.

Physics-based continuum modeling is the most useful and widely-utilized method for simulating the cell-level performance of metal-air batteries. Models constructed with this method are able to give researchers insight into a range of phenomena including the coupled effects of electrolyte concentrations, precipitation of solids, electrode kinetics, and cell voltage. The versatility of continuum modeling and the wide array of existing literature on the subject make it a good tool to advance the development of metal-air batteries.

Numerical modeling and simulation studies have shown that the performance LABs with non-aqueous electrolytes is encouraging but limited by slow oxygen transport and pore blockage by Li_2O_2. DFT simulations have been applied to elucidate the electronic structure and reaction mechanisms of Li_2O_2 and investigate possible alternative non-aqueous electrolyte solvents. Continuum models have highlighted the cell-level effects of Li_2O_2 precipitation and O_2 transport. The nucleation and growth of Li_2O_2 particles and films and its effect on the reversibility and performance of non-alkaline LABs is a major topic of research. Aqueous LABs improve oxygen transport in the air electrode and facilitate more favorable precipitation, but the long-term stability of the electrolyte is limited when exposed to CO_2. Continuum models developed for aqueous LABs have provided a framework for integrating the multi-phase flow in the air electrode into models of other metal-air systems.

ZABs stand alone as the only successfully commercialized primary metal-air system so far. Modeling studies of these systems highlight the performance of the alkaline electrolyte in air, passivation and shape change of the Zn electrode, and cell-level engineering. Because of its historical dominance, there is a long history of Zn-air continuum models, going back to the 1980s. Recent studies have been provided a scheme for interpreting and understanding experimental results, and a new framework developed to model ZABs with alternative near-neutral electrolytes could find wide application in other electrochemical systems. Implementing these modeling tools in the design process brings researchers closer to the goal of building high-performance and electrically rechargeable zinc-air batteries.

Acknowledgments: This work has received funding from the European Union's Horizon 2020 research and innovation program under grant agreement No. 646186 (ZAS-project). The support of the bwHPC initiative through the use of the JUSTUS HPC facility at Ulm University is acknowledged. The authors wish to thank Timo Danner for the helpful discussion.

Conflicts of Interest: The authors declare no conflict of interest.

References

1. Gröger, O.; Gasteiger, H.A.; Suchsland, J.P. Review—Electromobility: Batteries or Fuel Cells? *J. Electrochem. Soc.* **2015**, *162*, A2605–A2622.
2. Rahman, M.A.; Wang, X.; Wen, C. High Energy Density Metal-Air Batteries: A Review. *J. Electrochem. Soc.* **2013**, *160*, A1759–A1771.
3. Clark, S.; Latz, A.; Horstmann, B. Rational Development of Neutral Aqueous Electrolytes for Zinc-Air Batteries. *ChemSusChem* **2017**, *10*, 4735–4747.
4. Li, Y.; Zhang, X.; Li, H.B.; Yoo, H.D.; Chi, X.; An, Q.; Liu, J.; Yu, M.; Wang, W.; Yao, Y. Mixed-phase mullite electrocatalyst for pH-neutral oxygen reduction in magnesium-air batteries. *Nano Energy* **2016**, *27*, 8–16.
5. Li, C.S.; Sun, Y.; Gebert, F.; Chou, S.L. Current Progress on Rechargeable Magnesium-Air Battery. *Adv. Energy Mater.* **2017**, *7*, doi:10.1002/aenm.201700869.
6. Chen, L.D.; Norskov, J.K.; Luntz, A.C. Theoretical Limits to the Anode Potential in Aqueous Mg-Air Batteries. *J. Phys. Chem. C* **2015**, *119*, 19660–19667.
7. Vardar, G.; Sleightholme, A.E.S.; Naruse, J.; Hiramatsu, H.; Siegel, D.J.; Monroe, C.W. Electrochemistry of magnesium electrolytes in ionic liquids for secondary batteries. *ACS Appl. Mater. Interfaces* **2014**, *6*, 18033–18039.
8. Mokhtar, M.; Talib, M.Z.M.; Majlan, E.H.; Tasirin, S.M.; Ramli, W.M.F.W.; Daud, W.R.W.; Sahari, J. Recent developments in materials for aluminum-air batteries: A review. *J. Ind. Eng. Chem.* **2015**, *32*, 1–20.
9. Zhang, Z.; Zuo, C.; Liu, Z.; Yu, Y.; Zuo, Y.; Song, Y. All-solid-state Al-air batteries with polymer alkaline gel electrolyte. *J. Power Sources* **2014**, *251*, 470–475.

10. Fu, G.; Chen, Y.; Cui, Z.; Li, Y.; Zhou, W.; Xin, S.; Tang, Y.; Goodenough, J.B. Novel Hydrogel-Derived Bifunctional Oxygen Electrocatalyst for Rechargeable Air Cathodes. *Nano Lett.* **2016**, *16*, 6516–6522.
11. Adelhelm, P.; Hartmann, P.; Bender, C.L.; Busche, M.; Eufinger, C.; Janek, J. From lithium to sodium: Cell chemistry of room temperature sodium-air and sodium-sulfur batteries. *Beilstein J. Nanotechnol.* **2015**, *6*, 1016–1055.
12. Hartmann, P.; Bender, C.L.; Vračar, M.; Dürr, A.K.; Garsuch, A.; Janek, J.; Adelhelm, P. A rechargeable room-temperature sodium superoxide (NaO_2) battery. *Nat. Mater.* **2012**, *12*, 228–232.
13. Kim, J.; Park, H.; Lee, B.; Seong, W.M.; Lim, H.D.; Bae, Y.; Kim, H.; Kim, W.K.; Ryu, K.H.; Kang, K. Dissolution and ionization of sodium superoxide in sodium–oxygen batteries. *Nat. Commun.* **2016**, *7*, 10670, doi:10.1038/ncomms10670.
14. Durmus, Y.E. Modeling of Silicon Air-Batteries. Master's Thesi, Ulm University, Ulm, Germany, 2013, p. 106.
15. Cohn, G.; Ein-Eli, Y. Study and development of non-aqueous silicon-air battery. *J. Power Sources* **2010**, *195*, 4963–4970.
16. Durmus, Y.E.; Aslanbas, O.; Kayser, S.; Tempel, H.; Hausen, F.; de Haart, L.G.; Granwehr, J.; Ein-Eli, Y.; Eichel, R.A.; Kungl, H. Long run discharge, performance and efficiency of primary Silicon–air cells with alkaline electrolyte. *Electrochim. Acta* **2017**, *225*, 215–224.
17. Girishkumar, G.; McCloskey, B.; Luntz, A.C.; Swanson, S.; Wilcke, W. Lithium-air battery: Promise and challenges. *J. Phys. Chem. Lett.* **2010**, *1*, 2193–2203.
18. Lee, J.S.; Kim, S.T.; Cao, R.; Choi, N.S.; Liu, M.; Lee, K.T.; Cho, J. Metal-air batteries with high energy density: Li-air versus Zn-air. *Adv. Energy Mater.* **2011**, *1*, 34–50.
19. Li, Y.; Dai, H. Recent advances in zinc–air batteries. *Chem. Soc. Rev.* **2014**, *43*, 5257–5275.
20. Blurton, K.F.; Sammells, A.F. Metal/air batteries: Their status and potential—A review. *J. Power Sources* **1979**, *4*, 263–279.
21. Abraham, K.M.; Jiang, Z. A Polymer Electrolyte-Based Rechargeable Lithium/Oxygen Battery. *J. Electrochem. Soc.* **1996**, *143*, 1–5.
22. Xu, W.; Xu, K.; Viswanathan, V.V.; Towne, S.A.; Hardy, J.S.; Xiao, J.; Nie, Z.; Hu, D.; Wang, D.; Zhang, J.G. Reaction mechanisms for the limited reversibility of Li-O_2 chemistry in organic carbonate electrolytes. *J. Power Sources* **2011**, *196*, 9631–9639.
23. McCloskey, B.D.; Bethune, D.S.; Shelby, R.M.; Girishkumar, G.; Luntz, A.C. Solvents critical role in nonaqueous Lithium-Oxygen battery electrochemistry. *J. Phys. Chem. Lett.* **2011**, *2*, 1161–1166.
24. Freunberger, S.A.; Chen, Y.; Peng, Z.; Griffin, J.M.; Hardwick, L.J.; Bardé, F.; Novák, P.; Bruce, P.G. Reactions in the rechargeable lithium-O_2 battery with alkyl carbonate electrolytes. *J. Am. Chem. Soc.* **2011**, *133*, 8040–8047.
25. McCloskey, B.D.; Speidel, A.; Scheffler, R.; Miller, D.C.; Viswanathan, V.; Hummelshøj, J.S.; Nørskov, J.K.; Luntz, A.C. Twin problems of interfacial carbonate formation in nonaqueous Li-O_2 batteries. *J. Phys. Chem. Lett.* **2012**, *3*, 997–1001.
26. Albertus, P.; Girishkumar, G.; McCloskey, B.; Sanchez-Carrera, R.S.; Kozinsky, B.; Christensen, J.; Luntz, A.C. Identifying Capacity Limitations in the Li/Oxygen Battery Using Experiments and Modeling. *J. Electrochem. Soc.* **2011**, *158*, A343–A351.
27. Monroe, C.W. Does Oxygen Transport Affect the Cell Voltages of Metal/Air Batteries? *J. Electrochem. Soc.* **2017**, *164*, E3547–E3551.
28. Visco, S.J.; Katz, B.D.; Nimon, Y.S.; De Jonghe, L.C. Protected Active Metal Electrode and Battery Cell Structures with Non-Aqueous Interlayer Architecture. U.S. Patent 7282295 B2, 16 October 2007.
29. Visco, S.J.; Nimon, Y.S. Active Metal/aqueous Electrochemical Cells and Systems. U.S. Patent 7645543 B2, 12 January 2010.
30. Horstmann, B.; Danner, T.; Bessler, W.G. Precipitation in aqueous lithium–oxygen batteries: A model-based analysis. *Energy Environ. Sci.* **2013**, *6*, 1299–1314.
31. Hummelshøj, J.S.; Luntz, A.C.; Norskov, J.K. Theoretical evidence for low kinetic overpotentials in Li-O_2 electrochemistry. *J. Chem. Phys.* **2013**, *138*, 034703, doi:10.1063/1.4773242.
32. Sandhu, S.S.; Brutchen, G.W.; Fellner, J.P. Lithium/air cell: Preliminary mathematical formulation and analysis. *J. Power Sources* **2007**, *170*, 196–209.
33. Zheng, J.P.; Liang, R.Y.; Hendrickson, M.; Plichta, E.J. Theoretical Energy Density of Li–Air Batteries. *J. Electrochem. Soc.* **2008**, *155*, A432–A437.

34. Sahapatsombut, U.; Cheng, H.; Scott, K. Modelling of operation of a lithium-air battery with ambient air and oxygen-selective membrane. *J. Power Sources* **2014**, *249*, 418–430.
35. Horstmann, B.; Gallant, B.; Mitchell, R.; Bessler, W.G.; Shao-Horn, Y.; Bazant, M.Z. Rate-dependent morphology of Li_2O_2 growth in Li-O_2 batteries. *J. Phys. Chem. Lett.* **2013**, *4*, 4217–4222.
36. Danner, T.; Horstmann, B.; Wittmaier, D.; Wagner, N.; Bessler, W.G. Reaction and transport in Ag/Ag_2O gas diffusion electrodes of aqueous Li-O_2 batteries: Experiments and modeling. *J. Power Sources* **2014**, *264*, 320–332.
37. Lu, J.; Jung Lee, Y.; Luo, X.; Chun Lau, K.; Asadi, M.; Wang, H.H.; Brombosz, S.; Wen, J.; Zhai, D.; Chen, Z.; et al. A lithium–oxygen battery based on lithium superoxide. *Nature* **2016**, *529*, 377–382.
38. Johnson, L.; Li, C.; Liu, Z.; Chen, Y.; Freunberger, S.A.; Ashok, P.C.; Praveen, B.B.; Dholakia, K.; Tarascon, J.M.; Bruce, P.G. Erratum: The role of LiO_2 solubility in O_2 reduction in aprotic solvents and its consequences for Li-O_2 batteries. *Nat. Chem.* **2014**, *7*, 87.
39. Gao, X.; Chen, Y.; Johnson, L.; Bruce, P.G. Erratum: Promoting solution phase discharge in Li-O_2 batteries containing weakly solvating electrolyte solutions. *Nat. Mater.* **2016**, *15*, 918.
40. McCloskey, B.D.; Bethune, D.S.; Shelby, R.M.; Mori, T.; Scheffler, R.; Speidel, A.; Sherwood, M.; Luntz, A.C. Limitations in rechargeability of Li-O_2 batteries and possible origins. *J. Phys. Chem. Lett.* **2012**, *3*, 3043–3047.
41. Vegge, T.; Garcia-Lastra, J.M.; Siegel, D.J. Lithium–oxygen batteries: At a crossroads? *Curr. Opin. Electrochem.* **2017**, *6*, 100–107.
42. Pei, P.; Wang, K.; Ma, Z. Technologies for extending zinc-air battery's cyclelife: A review. *Appl. Energy* **2014**, *128*, 315–324.
43. Thomas Goh, F.W.; Liu, Z.; Hor, T.S.A.; Zhang, J.; Ge, X.; Zong, Y.; Yu, A.; Khoo, W. A Near-Neutral Chloride Electrolyte for Electrically Rechargeable Zinc-Air Batteries. *J. Electrochem.Soc.* **2014**, *161*, A2080–A2086.
44. Sumboja, A.; Ge, X.; Zheng, G.; Goh, F.T.; Hor, T.A.; Zong, Y.; Liu, Z. Durable rechargeable zinc-air batteries with neutral electrolyte and manganese oxide catalyst. *J. Power Sources* **2016**, *332*, 330–336.
45. MacFarlane, D.R.; Tachikawa, N.; Forsyth, M.; Pringle, J.M.; Howlett, P.C.; Elliott, G.D.; Davis, J.H.; Watanabe, M.; Simon, P.; Angell, C.A. Energy applications of ionic liquids. *Energy Environ. Sci.* **2014**, *7*, 232–250.
46. Liu, Z.; Abedin, S.Z.E.; Endres, F. Electrodeposition of zinc films from ionic liquids and ionic liquid/water mixtures. *Electrochim. Acta* **2013**, *89*, 635–643.
47. Liu, Z.; El Abedin, S.Z.; Endres, F. Dissolution of zinc oxide in a protic ionic liquid with the 1-methylimidazolium cation and electrodeposition of zinc from ZnO/ionic liquid and ZnO/ionic liquid-water mixtures. *Electrochem. Commun.* **2015**, *58*, 46–50.
48. Fu, J.; Cano, Z.P.; Park, M.G.; Yu, A.; Fowler, M.; Chen, Z. Electrically Rechargeable Zinc–Air Batteries: Progress, Challenges, and Perspectives. *Adv. Mater.* **2017**, *29*, 1604685.
49. Amendola, S.; Binder, M.; Black, P.J.; Sharp-Goldman, S.; Johnson, L.; Kunz, M.; Oster, M.; Chciuk, T.; Johnson, R. Electrically Rechargeable, Metal-Air Battery Systems and Methods. U.S. Patent 2012/0021303 A1, 26 January 2012.
50. Friesen, C.; Krishnan, R.; Friesen, G. Rechargeable Electrochemical Cell System with a Charging Electrode Charge/discharge Mode Switching in the Cells. U.S. Patent US 2011/0070506 A1, 24 March 2011.
51. Lee, D.U.; Xu, P.; Cano, Z.P.; Kashkooli, A.G.; Park, M.G.; Chen, Z. Recent progress and perspectives on bi-functional oxygen electrocatalysts for advanced rechargeable metal–air batteries. *J. Mater. Chem. A* **2016**, *4*, 7107–7134.
52. Neburchilov, V.; Wang, H.; Martin, J.J.; Qu, W. A review on air cathodes for zinc-air fuel cells. *J. Power Sources* **2010**, *195*, 1271–1291.
53. Stoerzinger, K.a.; Risch, M.; Han, B.; Shao-Horn, Y. Recent Insights into Manganese Oxides in Catalyzing Oxygen Reduction Kinetics. *ACS Catal.* **2015**, *5*, 6021–6031.
54. Horn, Q.C.; Shao-Horn, Y. Morphology and Spatial Distribution of ZnO Formed in Discharged Alkaline Zn/MnO_2 AA Cells. *J. Electrochem. Soc.* **2003**, *150*, A652–A658.
55. Stamm, J.; Varzi, A.; Latz, A.; Horstmann, B. Modeling nucleation and growth of zinc oxide during discharge of primary zinc-air batteries. *J. Power Sources* **2017**, *360*, 136–149.
56. Jörissen, L. Bifunctional oxygen/air electrodes. *J. Power Sources* **2006**, *155*, 23–32.
57. Drillet, J.F.; Holzer, F.; Kallis, T.; Müller, S.; Schmidt, V. Influence of CO_2 on the stability of bifunctional oxygen electrodes for rechargeable zinc/air batteries and study of different CO_2 filter materials. *Phys. Chem. Chem. Phys.* **2001**, *3*, 368–371.

58. Hwang, B.; Oh, E.S.; Kim, K. Observation of electrochemical reactions at Zn electrodes in Zn-air secondary batteries. *Electrochim. Acta* **2016**, *216*, 484–489.
59. Suresh Kannan, A.R.; Muralidharan, S.; Sarangapani, K.B.; Balaramachandran, V.; Kapali, V. Corrosion and anodic behaviour of zinc and its ternary alloys in alkaline battery electrolytes. *J. Power Sources* **1995**, *57*, 93–98.
60. Vorkapicc, L.Z.; Drazzicc, D.M.; Despic, A.R. Corrosion of Pure and Amalgamated Zinc in Concentrated Alkali Hydroxide Solutions. *J. Electrochem. Soc.* **1974**, *121*, 1385–1392
61. Baugh, L.M.; Tye, F.L.; White, N.C. Corrosion and polarization characteristics of zinc in battery electrolyte analogues and the effect of amalgamation. *J. Appl. Electrochem.* **1983**, *13*, 623–635.
62. Dirkse, T.; Hampson, N. The anodic behaviour of zinc in aqueous solution—III. Passivation in mixed KF-KOH solutions. *Electrochim. Acta* **1972**, *17*, 813–818.
63. Schröder, D.; Sinai Borker, N.N.; König, M.; Krewer, U. Performance of zinc air batteries with added K_2CO_3 in the alkaline electrolyte. *J. Appl. Electrochem.* **2015**, *45*, 427–437.
64. Mainar, A.R.; Leonet, O.; Bengoechea, M.; Boyano, I.; de Meatza, I.; Kvasha, A.; Guerfi, A.; Blazquez, J.A. Alkaline aqueous electrolytes for secondary zinc-air batteries: An overview. *Int. J. Energy Res.* **2016**, *40*, 1032–1049.
65. Mainar, A.R.; Iruin, E.; Colmenares, L.C.; Kvasha, A.; de Meatza, I.; Bengoechea, M.; Leonet, O.; Boyano, I.; Zhang, Z.; Blazquez, J.A. An overview of progress in electrolytes for secondary zinc-air batteries and other storage systems based on zinc. *J. Energy Storage* **2018**, *15*, 304–328.
66. Kar, M.; Winther-Jensen, B.; Forsyth, M.; MacFarlane, D.R. Chelating ionic liquids for reversible zinc electrochemistry. *Phys. Chem. Chem. Phys.* **2013**, *15*, 7191–7197.
67. Lee, J.S.; Lee, T.; Song, H.K.; Cho, J.; Kim, B.S. Ionic liquid modified graphene nanosheets anchoring manganese oxide nanoparticles as efficient electrocatalysts for Zn–air batteries. *Energy Environ. Sci.* **2011**, *4*, 4148–4154
68. Endres, F.; Abbott, A.; MacFarlane, D.R. (Eds) *Electrodeposition from Ionic Liquids*, 1st ed.; Wiley-VCH: Weinheim, Germany, 2008; p. 410.
69. Xu, M.; Ivey, D.G.; Xie, Z.; Qu, W. Rechargeable Zn-air batteries: Progress in electrolyte development and cell configuration advancement. *J. Power Sources* **2015**, *283*, 358–371.
70. Mclarnon, F.R.; Cairns, E.J. The Secondary Alkaline Zinc Electrode. *J. Electrochem. Soc.* **1991**, *138*, 645–664.
71. Kim, H.; Jeong, G.; Kim, Y.U.; Kim, J.H.; Park, C.M.; Sohn, H.J. Metallic anodes for next generation secondary batteries. *Chem. Soc. Rev.* **2013**, *42*, 9011–9034.
72. Bockris, J.O.; Nagy, Z.; Damjanovic, A. On the Deposition and Dissolution of Zinc in Alkaline Solutions. *J. Electrochem. Soc.* **1972**, *119*, 285–295.
73. Sun, K.E.; Hoang, T.K.; Doan, T.N.L.; Yu, Y.; Zhu, X.; Tian, Y.; Chen, P. Suppression of Dendrite Formation and Corrosion on Zinc Anode of Secondary Aqueous Batteries. *ACS Appl. Mater. Interfaces* **2017**, *9*, 9681–9687.
74. Banik, S.J.; Akolkar, R. Suppressing Dendritic Growth during Alkaline Zinc Electrodeposition using Polyethylenimine Additive. *Electrochim. Acta* **2015**, *179*, 475–481.
75. Diggle, J.W.; Damjanovic, A. The inhibition of the dendritic electrocrystallization of zinc from doped alkaline zincate solutions. *J. Electrochem. Soc.* **1972**, *119*, 1649–1658.
76. Schmid, M.; Willert-Porada, M. Electrochemical behavior of zinc particles with silica based coatings as anode material for zinc air batteries with improved discharge capacity. *J. Power Sources* **2017**, *351*, 115–122.
77. Zelger, C.; Laumen, J.; Laskos, A.; Gollas, B. Rota-Hull Cell Study on Pulse Current Zinc Electrodeposition from Alkaline Electrolytes. *Electrochim. Acta* **2016**, *213*, 208–216.
78. Liu, M.B. Passivation of Zinc Anodes in KOH Electrolytes. *J. Electrochem. Soc.* **1981**, *128*, 1663–1668.
79. Dirkse, T.; Hampson, N. The anodic behaviour of zinc in aqueous KOH solution—II. passivation experiments using linear sweep voltammetry. *Electrochim. Acta* **1972**, *17*, 387–394.
80. Zhang, X.G. *Corrosion and Electrochemistry of Zinc*, 1st ed.; Plenum Press: New York, NY, USA, 1996.
81. Prentice, G. A Model for the Passivation of the Zinc Electrode in Alkaline Electrolyte. *J. Electrochem. Soc.* **1991**, *138*, 890–894.
82. Jung, J.I.; Risch, M.; Park, S.; Kim, M.G.; Nam, G.; Jeong, H.Y.; Shao-Horn, Y.; Cho, J. Optimizing nanoparticle perovskite for bifunctional oxygen electrocatalysis. *Energy Environ. Sci.* **2016**, *9*, 176–183.
83. Mainar, A.R.; Colmenares, L.C.; Leonet, O.; Alcaide, F.; Iruin, J.J.; Weinberger, S.; Hacker, V.; Iruin, E.; Urdanpilleta, I.; Blazquez, J.A. Manganese oxide catalysts for secondary zinc air batteries: From electrocatalytic activity to bifunctional air electrode performance. *Electrochim. Acta* **2016**, *217*, 80–91.

84. Li, Y.; Gong, M.; Liang, Y.; Feng, J.; Kim, J.E.; Wang, H.; Hong, G.; Zhang, B.; Dai, H. Advanced zinc-air batteries based on high-performance hybrid electrocatalysts. *Nat. Commun.* **2013**, *4*, 1805.
85. Tan, P.; Kong, W.; Shao, Z.; Liu, M.; Ni, M. Advances in modeling and simulation of Li–air batteries. *Prog. Energy Combust. Sci.* **2017**, *62*, 155–189.
86. Grew, K.N.; Chiu, W.K. A review of modeling and simulation techniques across the length scales for the solid oxide fuel cell. *J. Power Sources* **2012**, *199*, 1–13.
87. Sholl, D. S.; Steckel, J.A. *Density Functional Theory: A Practical Introduction*, 1st ed.; John Wiley & Sons: Hoboken, NJ, USA, 2009.
88. Mattsson, A.E.; Schultz, P.A.; Desjarlais, M.P.; Mattsson, T.R.; Leung, K. Designing meaningful density functional theory calculations in materials science—A primer. *Model. Simul. Mater. Sci. Eng.* **2005**, *13*, 1–31.
89. Hörmann, N.G.; Jäckle, M.; Gossenberger, F.; Roman, T.; Forster-Tonigold, K.; Naderian, M.; Sakong, S.; Groß, A. Some challenges in the first-principles modeling of structures and processes in electrochemical energy storage and transfer. *J. Power Sources* **2015**, *275*, 531–538.
90. Nørskov, J.K.; Rossmeisl, J.; Logadottir, A.; Lindqvist, L.; Kitchin, J.R.; Bligaard, T.; Jónsson, H. Origin of the overpotential for oxygen reduction at a fuel-cell cathode. *J. Phys. Chem. B* **2004**, *108*, 17886–17892.
91. Viswanathan, V.; Hansen, H.A.; Rossmeisl, J.; Nørskov, J.K. Universality in oxygen reduction electrocatalysis on metal surfaces. *ACS Catal.* **2012**, *2*, 1654–1660.
92. Hansen, H.A.; Viswanathan, V.; Nørskov, J.K. Unifying kinetic and thermodynamic analysis of 2 e- and 4 e-reduction of oxygen on metal surfaces. *J. Phys. Chem. C* **2014**, *118*, 6706–6718.
93. Viswanathan, V.; Nørskov, J.K.; Speidel, A.; Scheffler, R.; Gowda, S.; Luntz, A.C. Li-O_2 kinetic overpotentials: Tafel plots from experiment and first-principles theory. *J. Phys. Chem. Lett.* **2013**, *4*, 556–560.
94. Aetukuri, N.B.; McCloskey, B.D.; Garcia, J.M.; Krupp, L.E.; Viswanathan, V.; Luntz, A.C. Solvating additives drive solution-mediated electrochemistry and enhance toroid growth in non-aqueous Li-O_2 batteries. *Nat. Chem.* **2015**, *7*, 50–56.
95. Eberle, D.; Horstmann, B. Oxygen Reduction on Pt(111) in Aqueous Electrolyte: Elementary Kinetic Modeling. *Electrochim. Acta* **2014**, *137*, 714–720.
96. Tripkovic, V.; Vegge, T. Potential and Rate Determining Step for Oxygen Reduction on Pt(111). *J. Phys. Chem. C* **2017**, *121*, 26785–26793.
97. Vazquez-Arenas, J.; Ramos-Sanchez, G.; Franco, A.A. A multi-scale model of the oxygen reduction reaction on highly active graphene nanosheets in alkaline conditions. *J. Power Sources* **2016**, *328*, 492–502.
98. Siahrostami, S.; Tripković, V.; Lundgaard, K.T.; Jensen, K.E.; Hansen, H.A.; Hummelshøj, J.S.; Mýrdal, J.S.G.; Vegge, T.; Nørskov, J.K.; Rossmeisl, J. First principles investigation of zinc-anode dissolution in zinc–air batteries. *Phys. Chem. Chem. Phys.* **2013**, *15*, 6416-6421
99. Jäckle, M.; Groß, A. Microscopic properties of lithium, sodium, and magnesium battery anode materials related to possible dendrite growth. *J. Chem. Phys.* **2016**, *141*, doi:10.1063/1.4901055.
100. Hummelshøj, J.S.; Blomqvist, J.; Datta, S.; Vegge, T.; Rossmeisl, J.; Thygesen, K.S.; Luntz, A.C.; Jacobsen, K.W.; Nørskov, J.K. Communications: Elementary oxygen electrode reactions in the aprotic Li-air battery. *J. Chem. Phys.* **2010**, *132*, doi:10.1063/1.3298994.
101. Radin, M.D.; Rodriguez, J.F.; Tian, F.; Siegel, D.J. Lithium peroxide surfaces are metallic, while lithium oxide surfaces are not. *J. Am. Chem. Soc.* **2012**, *134*, 1093–1103.
102. Radin, M.D.; Siegel, D.J. Charge transport in lithium peroxide: relevance for rechargeable metal–air batteries. *Energy Environ. Sci.* **2013**, *6*, 2370.
103. Radin, M.D.; Monroe, C.W.; Siegel, D.J. How dopants can enhance charge transport in Li_2O_2. *Chem. Mater.* **2015**, *27*, 839–847.
104. Dirkse, T.P. The Nature of the Zinc-Containing Ion in Strongly Alkaline Solutions. *J. Electrochem. Soc.* **1954**, *101*, 328.
105. Larcin, J.; Maskell, W.C.; Tye, F.L. Leclanche cell investigations I: $Zn(NH_3)_2Cl_2$ solubility and the formation of $ZnCl_2 \cdot 4\,Zn(OH)_2 \cdot H_2O$. *Electrochim. Acta* **1997**, *42*, 2649–2658.
106. Zhang, Y.; Muhammed, M. Critical evaluation of thermodynamics of complex formation of metal ions in aqueous solutions—VI. Hydrolysis and hydroxo-complexes of Zn2+ at 298.15 K. *Hydrometallurgy* **2001**, *60*, 215–236.
107. Clark, S.; Latz, A.; Horstmann, B. Cover Feature: Rational Development of Neutral Aqueous Electrolytes for Zinc-Air Batteries. *ChemSusChem* **2017**, *10*, 4666.

108. Smith, R.; Martell, A. *Critical Stability Constants*; Springer: New York, NY, USA, 1976; Volume 4.
109. Martell, A.E.; Smith, R.M. *Critical Stability Constants*; Springer: New York, NY, USA, 1977; Volume 5.
110. Martell, A.; Smith, R. *Other Organic Ligands*, 1st ed.; Springer: New York, NY, USA, 1977.
111. Dirkse, T. The Solubility Product Constant of ZnO. *J. Electrochem. Soc.* **1986**, *133*, 1656–1657
112. Limpo, J.; Luis, A. Solubility of zinc chloride in ammoniacal ammonium chloride solutions. *Hydrometallurgy* **1993**, *32*, 247–260.
113. Limpo, J.; Luis, A.; Cristina, M. Hydrolysis of zinc chloride in aqueous ammoniacal ammonium chloride solutions. *Hydrometallurgy* **1995**, *38*, 235–243.
114. Vazquez-Arenas, J.; Sosa-Rodriguez, F.; Lazaro, I.; Cruz, R. Thermodynamic and electrochemistry analysis of the zinc electrodeposition in NH_4Cl-NH_3 electrolytes on Ti, Glassy Carbon and 316L Stainless Steel. *Electrochim. Acta* **2012**, *79*, 109–116.
115. Rojas-Hernández, A.; Ramírez, M.T.; Ibáñez, J.G.; González, I. Construction of multicomponent Pourbaix diagrams using generalized species. *J. Electrochem. Soc.* **1991**, *138*, 365–371.
116. Newman, J.; Thmoas-Alyea, K.E. *Electrochemical Systems*, 3rd ed.; John Wiley & Sons: Hoboken, NJ, USA, 2004.
117. Husch, T.; Korth, M. Charting the known chemical space for non-aqueous lithium–air battery electrolyte solvents. *Phys. Chem. Chem. Phys.* **2015**, *17*, 22596–22603.
118. Chen, S.; Doolen, G.D. Lattice Boltzmann Method for Fluid Flows. *Annu. Rev. Fluid Mech.* **1998**, *30*, 329–364.
119. Trouette, B. Lattice Boltzmann simulations of a time-dependent natural convection problem. *Comput. Math. Appl.* **2013**, *66*, 1360–1371.
120. Aidun, C.K.; Clausen, J.R. Lattice-Boltzmann Method for Complex Flows. *Annu. Rev. Fluid Mech.* **2010**, *42*, 439–472.
121. Sukop, M.C.; Thorne, D.T. *Lattice Boltzmann Modeling*; Springer: Berlin, Germany, 2006; Volume 79, p. 016702.
122. He, X.; Li, N. Lattice Boltzmann simulation of electrochemical systems. *Comput. Phys. Commun.* **2000**, *129*, 158–166.
123. Niu, X.D.; Munekata, T.; Hyodo, S.A.; Suga, K. An investigation of water-gas transport processes in the gas-diffusion-layer of a PEM fuel cell by a multiphase multiple-relaxation-time lattice Boltzmann model. *J. Power Sources* **2007**, *172*, 542–552.
124. Hao, L.; Cheng, P. Lattice Boltzmann simulations of water transport in gas diffusion layer of a polymer electrolyte membrane fuel cell. *J. Power Sources* **2010**, *195*, 3870–3881.
125. Molaeimanesh, G.R.; Akbari, M.H. A three-dimensional pore-scale model of the cathode electrode in polymer-electrolyte membrane fuel cell by lattice Boltzmann method. *J. Power Sources* **2014**, *258*, 89–97.
126. Kim, K.N.; Kang, J.H.; Lee, S.G.; Nam, J.H.; Kim, C.J. Lattice Boltzmann simulation of liquid water transport in microporous and gas diffusion layers of polymer electrolyte membrane fuel cells. *J. Power Sources* **2015**, *278*, 703–717.
127. Ostadi, H.; Rama, P.; Liu, Y.; Chen, R.; Zhang, X.X.; Jiang, K. 3D reconstruction of a gas diffusion layer and a microporous layer. *J. Membr. Sci.* **2010**, *351*, 69–74.
128. Danner, T.; Eswara, S.; Schulz, V.P.; Latz, A. Characterization of gas diffusion electrodes for metal-air batteries. *J. Power Sources* **2016**, *324*, 646–656.
129. Sunu, W.G.; Bennion, D.N. Transient and Failure Analyses of the Porous Zinc Electrode. *J. Electrochem. Soc.* **1980**, *127*, 2007–2016.
130. Bockelmann, M.; Reining, L.; Kunz, U.; Turek, T. Electrochemical characterization and mathematical modeling of zinc passivation in alkaline solutions: A review. *Electrochim. Acta* **2017**, *237*, 276–298.
131. Neidhardt, J.P.; Fronczek, D.N.; Jahnke, T.; Danner, T.; Horstmann, B.; Bessler, W.G. A Flexible Framework for Modeling Multiple Solid, Liquid and Gaseous Phases in Batteries and Fuel Cells. *J. Electrochem. Soc.* **2012**, *159*, A1528–A1542.
132. Hein, S.; Latz, A. Influence of local lithium metal deposition in 3D microstructures on local and global behavior of Lithium-ion batteries. *Electrochim. Acta* **2016**, *201*, 354–365.
133. Hein, S.; Feinauer, J.; Westhoff, D.; Manke, I.; Schmidt, V.; Latz, A. Stochastic microstructure modeling and electrochemical simulation of lithium-ion cell anodes in 3D. *J. Power Sources* **2016**, *336*, 161–171.
134. Latz, A.; Zausch, J. Thermodynamic consistent transport theory of Li-ion batteries. *J. Power Sources* **2011**, *196*, 3296–3302.

135. Mao, Z.; White, R.E. Mathematical modeling of a primary zinc/air battery. *J. Electrochem. Soc.* **1992**, *139*, 1105–1114.
136. Deiss, E.; Holzer, F.; Haas, O. Modeling of an electrically rechargeable alkaline Zn-air battery. *Electrochim. Acta* **2002**, *47*, 3995–4010.
137. Andrei, P.; Zheng, J.P.; Hendrickson, M.; Plichta, E.J. Some Possible Approaches for Improving the Energy Density of Li-Air Batteries. *J. Electrochem. Soc.* **2010**, *157*, A1287–A1295
138. Schröder, D.; Krewer, U. Model based quantification of air-composition impact on secondary zinc air batteries. *Electrochim. Acta* **2014**, *117*, 541–553.
139. Arlt, T.; Schröder, D.; Krewer, U.; Manke, I. In operando monitoring of the state of charge and species distribution in zinc air batteries using X-ray tomography and model-based simulations. *Phys. Chem. Chem. Phys.* **2014**, *16*, 22273–22280.
140. Xue, K.H.; Nguyen, T.K.; Franco, A.A. Impact of the Cathode Microstructure on the Discharge Performance of Lithium Air Batteries: A Multiscale Model. *J. Electrochem. Soc.* **2014**, *161*, E3028–E3035.
141. Grübl, D.; Bessler, W.G. Cell design concepts for aqueous lithium-oxygen batteries: A model-based assessment. *J. Power Sources* **2015**, *297*, 481–491.
142. Yin, Y.; Gaya, C.; Torayev, A.; Thangavel, V.; Franco, A.A. Impact of Li_2O_2 Particle Size on Li-O_2 Battery Charge Process: Insights from a Multiscale Modeling Perspective. *J. Phys. Chem. Lett.* **2016**, *7*, 3897–3902.
143. Bazant, M.Z. Theory of chemical kinetics and charge transfer based on nonequilibrium thermodynamics. *Acc. Chem. Res.* **2013**, *46*, 1144–1160.
144. Latz, A.; Zausch, J. Thermodynamic derivation of a Butler-Volmer model for intercalation in Li-ion batteries. *Electrochim. Acta* **2013**, *110*, 358–362.
145. Breiter, M. Dissolution and passivation of vertical porous zinc electrodes in alkaline solution. *Electrochim. Acta* **1970**, *15*, 1297–1304.
146. Bai, P.; Bazant, M.Z. Charge transfer kinetics at the solid-solid interface in porous electrodes. *Nat. Commun.* **2014**, *5*, 1–13.
147. Zeng, Y.; Smith, R.B.; Bai, P.; Bazant, M.Z. Simple formula for Marcus-Hush-Chidsey kinetics. *J. Electroanal. Chem.* **2014**, *735*, 77–83.
148. Zeng, Y.; Bai, P.; Smith, R.B.; Bazant, M.Z. Simple formula for asymmetric Marcus-Hush kinetics. *J. Electroanal. Chem.* **2015**, *748*, 52–57.
149. Melander, M.; Jónsson, E.O.; Mortensen, J.J.; Vegge, T.; García Lastra, J.M. Implementation of Constrained DFT for Computing Charge Transfer Rates within the Projector Augmented Wave Method. *J. Chem. Theory Comput.* **2016**, *12*, 5367–5378.
150. Yuan, J.; Yu, J.S.; Sunden, B. Review on mechanisms and continuum models of multi-phase transport phenomena in porous structures of non-aqueous Li-Air batteries. *J. Power Sources* **2015**, *278*, 352–369.
151. Andersen, C.P.; Hu, H.; Qiu, G.; Kalra, V.; Sun, Y. Pore-Scale Transport Resolved Model Incorporating Cathode Microstructure and Peroxide Growth in Lithium-Air Batteries. *J. Electrochem. Soc.* **2015**, *162*, A1135–A1145.
152. Monroe, C.W.; Delacourt, C. Continuum transport laws for locally non-neutral concentrated electrolytes. *Electrochim. Acta* **2013**, *114*, 649–657.
153. Latz, A.; Zausch, J. Multiscale modeling of lithium ion batteries: Thermal aspects. *Beilstein J. Nanotechnol.* **2015**, *6*, 987–1007.
154. Xue, K.H.; McTurk, E.; Johnson, L.; Bruce, P.G.; Franco, A.A. A Comprehensive Model for Non-Aqueous Lithium Air Batteries Involving Different Reaction Mechanisms. *J. Electrochem. Soc.* **2015**, *162*, A614–A621.
155. Grübl, D.; Bergner, B.; Schröder, D.; Janek, J.; Bessler, W.G. Multistep Reaction Mechanisms in Nonaqueous Lithium-Oxygen Batteries with Redox Mediator: A Model-Based Study. *J. Phys. Chem. C* **2016**, *120*, 24623–24636.

MDPI
St. Alban-Anlage 66
4052 Basel
Switzerland
Tel. +41 61 683 77 34
Fax +41 61 302 89 18
www.mdpi.com

Batteries Editorial Office
E-mail: batteries@mdpi.com
www.mdpi.com/journal/batteries

www.ingramcontent.com/pod-product-compliance
Lightning Source LLC
LaVergne TN
LVHW070626170726
843515LV00004B/188

* 9 7 8 3 0 3 6 5 1 2 4 4 0 *